THE BRAIN

WHAT EVERYONE NEEDS TO KNOW®

THE BRAIN

WHAT EVERYONE NEEDS TO KNOW®

GARY L. WENK

OXFORD
UNIVERSITY PRESS

OXFORD

UNIVERSITY PRESS

Oxford University Press is a department of the University of Oxford.
It furthers the University's objective of excellence in research, scholarship,
and education by publishing worldwide. Oxford is a registered
trademark of Oxford University Press in
the UK and certain other countries.

"What Everyone Needs to Know" is a registered trademark of
Oxford University Press.

Published in the United States of America by Oxford University Press
198 Madison Avenue, New York, NY 10016, United States of America.

Library of Congress Cataloging-in-Publication Data
Names: Wenk, Gary Lee, author.
Title: The brain : what everyone needs to know / Gary L. Wenk, PhD.
Description: New York, NY : Oxford University Press, [2017] |
Series: What everyone needs to know |
Includes bibliographical references and index.
Identifiers: LCCN 2016030741 (print) |LCCN 2016042204 (ebook) |
ISBN 9780190603397 (pbk. : alk. paper) |
ISBN 9780190603403 (hardcover : alk. paper) | ISBN 9780190603410 (UPDF) |
ISBN 9780190603427 (EPUB)
Subjects: LCSH: Brain. | Neuropsychology. | Brain—Psychophysiology.
Classification: LCC QP376 .W46 2017 (print) | LCC QP376 (ebook) |
DDC 612.8—dc23
LC record available at https://lccn.loc.gov/2016030741

1 3 5 7 9 8 6 4 2

Paperback printed by LSC Communications, United States of America
Hardback printed by Bridgeport National Bindery, Inc.,
United States of America

For Jane

CONTENTS

3   How do food and drugs influence my brain?   41

6 How does my brain accomplish so much?   118

PREFACE

My goals in this book are to provide the most accurate and up-to-date information possible, with the caveat that "facts" evolve and continually are being modified by new knowledge, and to present this information in a language and format accessible to a novice reader. The chapters provide answers to questions such as: How did the brain evolve? What is an emotion? What is a hallucination? How do you learn? How does your diet affect how you think and feel? What happens to your brain as you age?

Many authors begin with a discussion of neuroscience with the assumption that the reader needs to know all of the basic brain anatomy and chemistry before proceeding. I have chosen to place this chapter on neuroscience at the end of the book. In addition, I have tried to keep the jargon to a minimum and have included occasional reminders of the meanings of terms within each chapter. Should you wish to know more about some terms, I have included a Glossary at the end of the book. I have written the chapters so that they can be read in any order; I encourage you to begin with the topic that you find most interesting.

René Descartes speculated that the mind exists independent from the brain. Today, almost four centuries after Descartes's death, we still do not fully understand how your mind emerges from the electrical and chemical processes that occur in your brain. This book introduces you to the answers that have been obtained thus far. I sincerely hope that reading this book encourages you to learn more about the wonderful organ of the mind that lives in your head. You will discover that our current understanding of the brain is incomplete. You also will discover that there are many interesting and related topics that have not been discussed in this book. My goal has been to present what I believe everyone *needs to know about the brain,* rather than *everything that is known about the brain.* In addition, I have chosen to focus upon only a few of the most common neurological diseases or mental illnesses, such as Alzheimer's and Parkinson's disease and depression.

During every stage of the writing, the text benefited immeasurably from the brilliant editorial suggestions of my wife, Jane, who shaped my concept of my audience and how to reach them; she skillfully converted my jargon into intelligible prose and contributed additional concepts and topics that made the book into a more comprehensive discussion of the brain. I learned to trust her judgment and insight more than my own; if there is wisdom in my writing, it has evolved under her guidance. For more than 36 years, I have been blessed to share my life with this wonderfully patient and intelligent woman who has enriched my life in countless ways. This book is dedicated to Jane.

I am deeply indebted to my editor, Joan Bossert, for asking me to submit a book on this topic in the Oxford series on *What Everyone Needs To Know®.* Since that day, the clarity of her advice and her unswerving encouragement have been invaluable. This book's core derives from a course of lectures that I have given for the past 35 years to first-year psychology and biology majors. At first, I assumed that I was teaching students about the brain; ultimately, I realized that I was teaching them how to understand themselves.

INTRODUCTION

Your brain is in your head. At first glance, this seems like a dreadful place for your delicate brain to reside. Why is your brain located inside your head? It would be much safer hidden deep inside your chest. With very few exceptions, brains are always located at the front end of an animal's feeding "tube" or digestive system that extends from the mouth to the anus. Bugs, worms, fish, birds, reptiles, dogs, and humans are simple feeding tubes with a "brain" sitting right up front, usually near the eyes, ears, and nose; thus, making it possible to find food by sight, sound, or smell and then to organize behavior so that the front end of the feeding tube can get close enough to "taste" the food and check its safety before engulfing it. Once the food is in your feeding tube, its nutrients are absorbed and become available to the cells in the rest of your body. The calories from a meal are not distributed equitably around your body. Imagine that the meal you have just finished is worth $1; the various digestive components of your feeding tube spend nearly 70 cents of your meal, with the remaining 30 pennies alone available to your brain and body to spend on its daily needs. Your brain and other organs that allow you to reproduce and move around your environment (including your muscles and bones) spend about 22 of those remaining pennies. As you can see, very few pennies are left over to spend on the other tasks in your body. These expenditures give you some idea of the priorities—thinking, sex, and movement—that billions of years of evolution have established for your brain and body (see Figure 1).

Human brains use a lot of energy. Under normal circumstances, the brain primarily uses energy in the form of sugar: the equivalent of about 12 donuts every day! (Now you can understand why there are so many donut shops located along your morning drive to work.) Your body spends nearly a quarter of its food budget on just the brain; five times as much as most other mammals devote to their brains. Your brain uses most of this energy to organize your behavior to find food, avoid danger, and socialize with others in order to find a mate with whom to reproduce. You know one manifestation of this

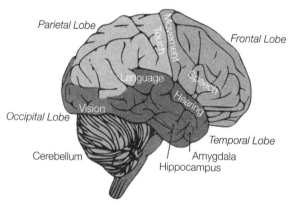

Figure 1 Your brain is the command and control center for your nervous system and body. It receives input from the sensory organs and sends output to the muscles. The human brain has the same basic structure as other mammal brains, but it is larger in relation to body size than any other mammal brain. The brain makes up about 2% of your body weight. The largest part of the human brain is the cerebrum, which is divided into two hemispheres. Underneath lies the brainstem, and behind that sits the cerebellum. The outermost layer of the cerebrum is the cerebral cortex, which consists of four lobes: the frontal lobe, the parietal lobe, the temporal lobe, and the occipital lobe. The cerebral cortex is responsible for complex thought, sensory processing, and movement. Visual processing takes place in the occipital lobe, near the back of the skull. The temporal lobe processes sound and language, and includes the hippocampus and amygdala, which play roles in memory and emotion, respectively. The parietal lobe integrates input from different senses and is important for spatial orientation and navigation.

imperative as dating, and it requires a very large and complex brain to pull this off successfully. Eating and sex are obviously excellent ideas if your purpose is to maintain and propagate your species; fortunately, your brain does an excellent job achieving both of these tasks.

The evolution of energy-guzzling bigger brains, like yours, depended on building longer feeding tubes in order to optimize the extraction of more energy from whatever entered the front end of the feeding tube. It is not surprising then that the length of the gut, when compared across many different species, correlates with the size of the brain. As brains became larger, however, the forces of evolution shifted strategies (after all, the length of the gut can be increased only until there is insufficient room in the body to contain it); animals developed a more efficient and shorter feeding tube that relied on a high-quality, nutrient-rich diet. Therefore, today we have a gastrointestinal system that is efficient at extracting energy for itself and its two principal customers, your reproductive system and brain. Due to the high energy demands of the brain and reproductive system, however, a surprising compromise occurred during evolution: as brains became bigger, human reproductive success failed. Now you can appreciate why humans do not give birth to litters. One might predict that having a larger brain would allow greater reproductive success. After all, you would expect that animals with bigger brains would find more food, avoid predators more successfully, and find more mates. This expectation is based on the assumption that bigger brains are always smarter, but this is not so. Animals with smaller brains and bodies, such as birds, often demonstrate impressive cognitive abilities, while some large-brained species, such as whales and elephants, do not.

Brains, regardless of how big or small, complex or simple, evolved to perform one basic function: survival of the individual and the species. Survival depends upon finding food and shelter, mating successfully, and avoiding predators; doing all of this requires the ability to learn how and where to hunt, to learn how to communicate, and to cooperate with others, while remembering what sounds and odors predict the imminent appearance of predators, as well as remembering whether it is better to fight or flee. The Spanish filmmaker Luis Buñuel famously stated that ". . . memory is what makes our lives. Life without memory is no life at all . . . Our memory is our coherence, our reason, our feeling, even our action. Without it we are nothing." Thus, I will begin by discussing how memories are made and lost.

1

WHAT IS A MEMORY?

Your brain is never static; it continually rewires itself as you experience life and learn new things. This ability is termed plasticity. Plasticity allows you to be smart, adaptive, and better able to survive in a changing and challenging environment. This same plasticity has a dark side; it underlies your tendency to become easily addicted to drugs, foods, sex, gambling, and potentially dangerous behaviors.

What does it mean to learn something? Learning is a relatively permanent change in behavior due to an increase in knowledge or skills in response to information stored in the brain. Different parts of the brain are responsible for different types of learning. Structures within the temporal lobe, a region of the brain near your ears, are responsible for learning about specific events in your life, such as what happened yesterday or on your birthday last year, and factual knowledge, such as $6 \times 7 = 42$ or "what is a chair?" These events and bits of specific information are called memories. Other regions of the brain store memories related to specific skills, such as how to swing a golf club or baseball bat, or how to ride a bicycle.

Memories are not simple recordings of isolated events or bits of knowledge; they contain aspects of time and space; for example, when and where something happened, as well as an emotional component that describes how you felt when the memory was formed. The process of storing different aspects of a memory in different brain regions dedicated to specific features of the memory, such as time or emotion, allows the human brain to be very good at storing lots of memories. Thus, the components of an event get distributed to various brain regions for long-term storage. The distributed nature of memories makes storing them more efficient and protects them from being totally lost due to injury or aging; however, the distributed nature of memories also makes retrieving these stored memories much more challenging. Why do the same mechanisms that make storing memories so easy underlie the difficulty we have in retrieving them? In order to answer this question, let us begin by examining how your brain makes memories.

How are memories made?

Your memory of your last birthday began as a complex mix of sensory events that included a large variety of odors, sights, tastes, sounds, and emotions; these experiences were first processed by the specific areas of your brain that are responsible for processing each of these sensory inputs. Your sensory experiences then were funneled through a structure called the hippocampus, which is located within the temporal lobe of the brain. This part of your brain lies near your ears. The hippocampus is responsible for binding together the diverse sensory elements required to create vivid and coherent memories full of emotion.

In addition, memories often get a specific time stamp associated with them. Let's use the example of the memory you have of your last birthday. First, the hippocampus gathers all of the sensory aspects of the event and conducts some initial processing of these sensory elements into a neurological format that is not yet completely understood. After this initial phase is completed, the entire memory of your last birthday is distributed widely throughout various brain regions. Sensory memories initially are stored in regions of cortex responsible for processing the particular type of sensation, for example, sight, sound, or odors. Thus, the components of a memory of an event in your life are stored throughout your brain.

Your memories are much more than just sensations. They also contain your feelings at the time (e.g., whether you were happy or sad). These are stored simultaneously in brain regions devoted to emotional memories (more on these brain regions in the next chapter). Ultimately, the complex sensory and emotional experience that was your birthday bash gets stored in many different brain regions; memories are not stored in just one place within the brain. By storing the information in many places your rich memory of your last birthday is far less likely to be lost due to injury or time. Due to the period of time that has passed since that birthday celebration, you might not be able to recall every aspect of your last birthday, but you always will be able to remember many of the most salient and emotional aspects of what happened.

We use our hippocampus to remember past events, familiar objects, people, and places; we also use it to construct possible futures. After all, how different is anyone's future from the past? For most of us, for most of our lives, tomorrow is often just a simple variation on yesterday and involves the same people and places. Some memories are stronger than others; indeed, some memories you might wish to forget. Strong memories usually have robust emotional components. On average, women retain more vivid emotional memories than men, particularly if those events have a strongly negative emotional component. Psychologists speculate that because women may tend to dwell on "memories of negative life experiences" more often than do men, women are diagnosed more often with depression. Memories that are consolidated in association with extremely unpleasant events can lead to post-traumatic stress disorder, a condition that is twice as common in women as in men.

Do I need to pay attention?

Memories form in a stepwise fashion in the brain. If you are paying attention to a particular sensory experience, usually because it has some importance to you at that moment, the collection of sensory inputs into your brain are temporarily stored as a short-term memory. Can you learn when you are not paying attention, or when you are bored or completely uninterested in what is happening around you? Yes, but not well, and certainly not efficiently, because these states produce interference with the memory consolidation process. The memory consolidation process usually occurs within the hippocampus and is rather unstable. It is easy for memories to be lost at this point by paying attention to other sensory inputs. We all have experienced the impact of distraction upon our ability to remember something. For example, most people find it difficult to concentrate or read a book in a noisy room. In contrast, if an experience is significant to you because it is associated with a strong emotion or has survival value, it is far more likely to be converted into a long-term memory that will last virtually indefinitely.

Why do I forget some things but not others?

Storing new memories, especially important memories, is easy, but the size of the brain is finite; it thus has a limited amount of storage space. You can easily grasp the problem: What to do with all of the information that flows into your brain every second of every day? The solution: delete the memories. Intentional forgetting, or deliberately erasing memories, plays just as vital a role in the brain as does remembering. The hippocampus automatically encodes all experience, yet the vast majority of our experiences are not remembered later. This forgetting is deliberate, but why does your brain intentionally forget things? Information bombards your senses every day, and your brain initially attempts to store this information just in case it is important. Subsequently, your brain wastes a considerable amount of storage space on useless information. For example, consider this scenario: Your roommate or spouse appears wearing a new shirt or blouse and you know immediately that it is new. How is this possible? Did you memorize his or her entire wardrobe? Apparently you did; otherwise you would not have noticed that the shirt or blouse is new. Clearly, we all waste a lot of brain space storing useless or unimportant information. As you will read later, our brain actively removes unwanted memories while we are sleeping; this process allows us to be smarter, and to learn information more critical to our survival, during the day.

How are memories recalled?

Your brain evolved to pay attention to everything that was novel because such knowledge might increase your chances of survival. Just because you have a specific memory stored somewhere in your brain, however, does not guarantee that you will be able to access it when required. Retrieving a memory is an active process that requires the brain to reproduce some component of the original memory trace in order to recall all of the essential pieces of the memory. Scientists speculate that the process of recalling a memory involves the restoration of the pattern of neuronal activations that was present during the encoding process. Recollection requires finding all of the component parts of a memory and then recombining all, or almost all, of the pieces together into a whole memory. This is a challenging task that your brain must accomplish in just a few seconds.

The act of retrieving a memory appears to render the memory highly susceptible to modification. This is a critical feature of our brain: memories are not video/audio recordings of actual events. Memories are made of the pieces of the event or experience that we were paying attention to at the time. Thus, when you retrieve a memory, you are recalling all of its parts into consciousness. The process of recalling a memory leads to distortions resulting from the incorporation of misinformation into the memory. Then, when you are done with the memory, your brain stores it away again. What this means is that familiar memories are recalled, retold, embellished (intentionally or not), and then re-stored as the newly edited storyline. Usually, you are not aware that this distortion occurred! Each time a memory is recalled it is vulnerable to alteration.

Do memories last forever?

Unfortunately, no, they do not last forever, at least not perfectly. The manner in which memories are recorded and stored makes certain aspects of them quite unstable over time. Lawyers often take advantage of this susceptibility of memories to alteration and distortion; it is easy for them to catch a witness in an apparent lie by comparing the witness's memories recorded immediately after an event with his or her recollection many months or years later. Test this out with your friends; ask them to remember where they were, what they were doing, or who they were with when they saw the Twin Towers in New York City collapse, or when they witnessed the space shuttle Challenger explode, or even (for you older readers) what they were doing when they learned that President Kennedy had been assassinated. Then compare their recollections with those of their friends who were (supposedly) with them during these tragic, emotionally charged events. You will quickly discover that they got their so-called facts wrong. They have retold and remembered the events surrounding these tragic days so many times that the original memory has been completely corrupted and altered. Our brains are not accurate recording devices; never depend upon them as such. Brains evolved to help you survive and procreate, not to record events with great detail. There was simply never any evolutionary pressure for our brains to record every detail of an event. The instability of making and recalling memories has numerous social and legal ramifications, such as the questionable value of eyewitness accounts, particularly for events that occurred in the distant past. Scientists recently have succeeded in implanting false memories in mice; that is, the mice behave as though they know something to be true even though the event never occurred. Social psychologists have demonstrated that humans are quite vulnerable to the implantation of false memories that are later recalled as "repressed memories" for experiences that never occurred. Your brain can very easily believe that specific events occurred, and discount all evidence to the contrary, and then fail to recall memories of events that you actually experienced. This is called amnesia.

What is amnesia?

Sometimes, old memories are not accessible for recall; this is called retrograde amnesia. Retrograde amnesia is memory loss for events that occurred before some kind of trauma or in response to degeneration of specific parts of the brain. Usually the amnesia is not comprehensive but only extends to events that occurred during the weeks or months before the incident. Because aspects of memories are stored in different brain regions, people with retrograde amnesia usually do not lose their entire memory store. Some aspects of the original memory are usually intact in these patients. For example, patients who have suffered a brain trauma that produced amnesia still remember many skills, such as how to walk, talk, and write, as well as many facts about the world, such as what a red light means or how to play with a yo-yo.

The inability to form new memories is called anterograde amnesia and is often due to brain trauma or to a neurodegenerative disease such as Alzheimer's disease. The symptoms can range from slowed learning to a complete inability to learn new things. Trauma or neurodegeneration usually results in the presence of both retrograde and anterograde amnesia with varying degrees of severity. For example, alcoholics usually spend all of their money on their addiction and, therefore, cannot afford to maintain a diet that contains all of the nutrients the brain needs to function; thus, a poor diet can lead to the degeneration of vulnerable brain regions. Ultimately, due to the degeneration of vulnerable regions in the temporal lobe and other nearby brain regions, alcoholics demonstrate both anterograde and retrograde amnesia in addition to mental confusion and personality changes. By studying how amnesia occurs, scientists have learned much about the biological mechanisms the brain uses to learn and remember.

How does the brain create a memory?

In order to understand how your brain makes a memory, you first need to learn about brain chemistry and the role specific chemicals play in the creation of a memory. First, you need to know about a chemical in the brain called acetylcholine. Acetylcholine is a neurotransmitter. A neurotransmitter is a chemical substance produced within neurons from components of the diet. There are many different neurotransmitters in the brain. They are released by neurons to diffuse into the extracellular environment in order to influence the behavior of nearby neurons. Acetylcholine exists almost everywhere in nature; it is not unique to your brain. Acetylcholine has been found in multicellular organisms as well as in blue-green algae, where it may be involved with photosynthesis. Acetylcholine stimulates silk production in spiders and limb regeneration in salamanders. In humans, acetylcholine enables movement by stimulating our muscles to contract; it also plays an important role in the actions of the autonomic nervous system.

The autonomic nervous system maintains homeostasis, or equilibrium, for your entire body. Among other functions, it controls the rate at which your heart beats, how fast you breathe, how much saliva your mouth produces, the rate of movement of material in your gut, your ability to initiate urination, how much you are perspiring, the size of your pupils, and the degree of visible sexual excitation you might experience. The actions of acetylcholine within your autonomic nervous system indirectly influence how you feel when memories are being recorded by your brain.

The human brain's numerous acetylcholine pathways influence the function of the cortex, hippocampus, and many other regions. Within these various regions, the actions of acetylcholine enable you to learn and remember, to regulate your attention and mood, and to control how well you move. Thus, anything that affects the function of acetylcholine has the potential to affect all of these brain and body functions. That "anything" could be a drug or disease.

Once released into the synapse, a small space where two neurons almost touch each other, the neurotransmitter acetylcholine can act on two quite different protein receptors: one receptor is named for muscarine and the other is named for nicotine. Most of the acetylcholine receptors in the brain respond to muscarine. Scientists know quite a lot about the role of these muscarine receptors because many plants contain chemicals that can block their normal function in the brain. Chemicals found in plants, such as henbane, Jimson weed, mandrake, and the Deadly Nightshade, selectively block the muscarine receptors; as a result, if you ingest these plants, you would quickly lose your ability to form new memories or pay attention to someone talking to you. All of these plants are in the *Solanaceae* family; another member of this family is the tobacco plant, which contains nicotine. Less than 10% of the acetylcholine receptors in your brain respond to nicotine. However, if those few nicotinic receptors did not exist in the brain, no one would bother smoking because the nicotine would not be psychoactive. I will discuss the impact of nicotine on the brain at the end of this chapter.

Acetylcholine does not act alone to make memories; it requires assistance from a very simple amino acid that is also an important neurotransmitter—glutamate. Glutamate makes and breaks connections between neurons and thus makes and breaks memories. It does this by allowing the passage of sodium or calcium ions into neurons. Following the entry of calcium ions, some truly interesting things begin to happen inside the neuron that leads to the production of a memory. Calcium ions activate a complex cascade of biochemical changes that ultimately involve the genes of the neuron and that may actually change how the neuron behaves for the rest of your life.

These biochemical changes also may alter how one neuron communicates with hundreds of other neurons throughout your brain. Think of this neural process as a symphony of musicians playing together for the first time. Initially, everyone is playing his or her own song. Finally, the conductor, that is, glutamate, arrives and hands out a musical score; all of the musicians begin to play a complex pattern of musical rhythms. In the same way that a pattern of sounds produced by a symphony conveys feeling, the rhythms of activity of the neurons in your brain convey information. Glutamate initiates the process of forming an ensemble of rhythms that is the basis of a memory. Your neurons, the individual cells that process thoughts and feelings, are the musicians and they become linked to one another according to a common pattern of activity. Scientists call this background rhythm that gets your neurons singing together the slow gamma rhythm. Once this linking occurs the neurons form a stable collaborative group of neural musicians that plays a particular song, or memory, which can recur only when that particular ensemble of neurons plays the same pattern of music together. In this analogy, memories can be seen as a unique song stored as a stable pattern of neural activity within your brain. Just as we enjoy playing the same tunes over and over again, we also enjoy replaying pleasant memories. You may know this song of neural activity as daydreaming. More on the importance of daydreaming later.

Glutamate often demonstrates quite different roles in your brain depending upon your age. When glutamate is functioning correctly, memories can be formed. When you are much older, or if you have Alzheimer's disease or have experienced a stroke, glutamate's behavior becomes destructive. When too much glutamate is present in the synapse, neurons may die and memories may be destroyed forever. Thus, maintaining a good balance of glutamate function is a challenging but critical requirement for neurons.

Glutamate also has a unique role in brain development. When you were an infant, the neurons in your brain developed many connections, or synapses, with other neurons to optimize your ability to learn a great amount of information quickly, such as how to move your hands and feet, the sound of your mother's voice, or what the color red looks like. But as you grew older (during early adolescence), your brain became a bit like an overwired computer—for it to work better and faster it became beneficial for it to remove unnecessary "wires," or connections. This is where glutamate's unique dual abilities come into play. Your brain used glutamate to break connections between neurons that had become unnecessary, which, in turn, allowed the remaining neural circuits to function more efficiently. Now, as an adult, glutamate allows your brain to be "plastic," to adjust your behavior to your environment in order to increase your chances of survival. Overall, the actions of glutamate are age-dependent: when you were young, it helped make thinking more efficient; as an adult, it is responsible for making memories last a lifetime. Sometimes, due to disease or degeneration, such as that associated with Alzheimer's disease, the brain has trouble making memories and may even lose the ones it has stored.

Why are patients with Alzheimer's disease so forgetful?

Neuroscientists have learned quite a lot about the role of acetylcholine and glutamate by investigating what happens when the neurons that release these neurotransmitters are injured or diseased. In the brains of people with Alzheimer's disease, acetylcholine neurons that project into the hippocampus and cortex slowly die. Glutamate's penchant for pruning that was so beneficial when the brain was young now becomes responsible for the death of acetylcholine neurons, as well as many other neurons, in the brains of patients with Alzheimer's disease. The loss of normal acetylcholine and glutamate function in the cortex may be why patients with Alzheimer's disease have difficulty paying attention to important events in their daily lives. The impaired function of acetylcholine and glutamate within the hippocampus may underlie the debilitating memory loss that is the earliest hallmark of this disease. The impaired function of these neurotransmitter systems in the brains of patients with Alzheimer's disease has led scientists to design treatments that might enhance the function of acetylcholine.

How is the memory loss treated?

Sometimes, the severity of the cognitive symptoms in Alzheimer's disease can be reduced, at least to some degree, by drugs and dietary nutrients that enhance the function of acetylcholine in the brain. To understand how this is possible, we need to look at how acetylcholine is produced in the brain. Neurons synthesize acetylcholine from choline, which is obtained from the diet, and from acetyl groups that originate in mitochondria from the metabolism of sugar. Sugar is a vital nutrient for your brain's normal function. The synthesis of acetylcholine occurs within the cytoplasm of your neurons and it is then released to communicate with neurons that are important for learning and memory to occur.

Many health food stores across America sell choline powder under the pretense that consuming more choline will enable the brain to make more acetylcholine. Given the vital role of acetylcholine in learning and memory, this is an alluring claim. Regrettably, it has no basis in fact. An important thing to realize is that the brain responds only to deficits, not surpluses, in the diet. The brain always has a ready source of choline from the diet (in donuts, cookies, cakes, eggs, beef, and fish) or from stores in the liver and, in fact, never develops a deficit in choline, even in patients with Alzheimer's disease.

Thus, consuming extra choline does not induce your brain to make more acetylcholine. Instead, it only results in a gaseous byproduct that you exhale and that smells like rotting fish. Rather than enhancing your cognitive abilities, choline supplements merely generate a terrible case of bad breath. Once released, the action of acetylcholine is terminated by an enzyme called acetylcholinesterase. Many different drugs are capable of inhibiting this enzyme, causing synaptic levels of acetylcholine to rise. Today, these drugs are given to patients with Alzheimer's disease to improve their ability to pay attention or remember the day's events. Although the benefits tend to be limited for most patients, neuroscientists are experimenting with better ways to enhance the action of acetylcholine and thus improve learning and memory abilities for patients with Alzheimer's disease.

It is also worth considering what would happen if a neuron could not release acetylcholine at all. The botulinum toxin released by the *Clostridium botulinum* bacteria that is sometimes found in the foods we eat can inhibit the release of acetylcholine from nerve terminals. Fortunately for your brain, this toxin cannot cross the blood–brain barrier. There is, however, more to you than just your brain. Botulinum toxin can significantly impair the ability of your vagus nerve to control your breathing. Your vagus nerve is responsible for causing the contraction of your diaphragm muscle; when this muscle contracts, it pulls air into your lungs. However, if your brain cannot communicate with your diaphragm via the release of acetylcholine from the vagus nerve, you will stop breathing and die. The botulinum toxin is exceptionally potent; 1 gram is sufficient to kill approximately 350,000 people!

When acetylcholine and glutamate are functioning normally, however, memories are easily made and stored. The sights, sounds, smells, tastes, and feel of life events are processed by the back half of the human brain; this information then is funneled into the temporal lobe where it becomes organized—primarily by the hippocampus—for long-term storage. We know from many investigations over the past few decades that memories are quite vulnerable to loss during this early stage of processing within the hippocampus. Once the memory is initially processed, it is transferred, usually while you are sleeping, to other brain regions for long-term storage. The most important thing to realize is that a single memory is not located in a single place in the brain; rather, various components of the memory are distributed throughout the brain.

The best evidence today indicates that a memory involves specific series of structural and biochemical modifications on both sides of the synapse. Simply stated, a memory involves making a change in the efficiency of neuronal connections. If you could miniaturize yourself within a brain, you would see that not every connection between neurons is equally efficient; there is typically quite a lot of noise and error in most neural circuits within the brain. In spite of all of this noise in the circuitry, however, your brain manages to store quite a lot of information.

What does a memory look like?

Memory-induced changes can be visualized by currently available techniques; they appear as structural changes in how two neurons form the synapse that connects them to each other. The structural changes often look like raised bumps on the surface of neurons. Learning leads to an increase in the number of these bumps; although they look a lot more like lollipops, they are called dendritic spines. Roughly speaking, bigger dendritic spines indicate stronger connections between neurons and stronger memories.

Studies have shown that people who extensively utilize their hippocampus, for example, very experienced London taxi cab drivers, actually have significantly bigger hippocampi than do cab drivers who are new on the job. Yes, as the cabby spent his days driving around London, his brain was busy growing and strengthening these connections between neurons in the hippocampus, which allowed him to form mental maps of the city. Drivers with better maps could get their passengers to their destinations faster and more efficiently, a process of path finding that was mirrored deep inside the cabby's hippocampus.

Animal studies have confirmed that the density of dendritic spines within the hippocampus also varies according to the stage of the menstrual cycle; the number of dendritic spines decreases after the surge of progesterone at the time of ovulation. If humans demonstrate a similar pattern of changes, then women are most likely to become pregnant when they are least able to remember the circumstances surrounding the event.

How does nicotine influence brain function?

Nicotine affects brain function in a dose-dependent fashion via its actions upon the brain's acetylcholine system. Low doses tend to activate your left hemisphere and produce mental stimulation and a feeling of arousal and attentiveness, whereas high doses tend to activate your right hemisphere more strongly and are closely associated with the sedative effects of nicotine. Therefore, when doing boring tasks, you could take a low dose of nicotine by, say, smoking one cigarette that would increase your subjective feelings of arousal and attention. In contrast, during anxious or stressful situations, you could take a high dose of nicotine by chain-smoking a few cigarettes and actually reduce your stress by activating the right hemisphere and producing a bit of sedation. These findings nicely demonstrate the competing roles of nicotine receptors in the two brain hemispheres and provide some insights into how the two halves of the brain normally function to produce a balance of emotions (in the right side of your brain) and attention and arousal (in the left side of your brain). Acetylcholine nicotinic receptors play an important role in attention; it is now known that

60% of adults diagnosed with attention deficit hyperactivity disorder (ADHD) smoke cigarettes as compared with less than 30% of the rest of the population. These adult ADHD patients are finding ways of self-medicating themselves to improve brain function while simultaneously increasing their risk of developing lung cancer.

Schizophrenia patients have fewer and more poorly functioning nicotinic receptors, especially in brain areas involved in the expression of their specific cognitive and sensory deficits. For example, the loss of function of these nicotinic receptors in their brains may contribute to impaired attentional abilities and memory. The important role of nicotinic receptors in schizophrenia was inferred based on patients' distinctive cigarette-smoking habits. Ninety percent of schizophrenic patients smoke; they smoke more cigarettes per day, inhale more deeply, and smoke their cigarettes to the butt more often than non-schizophrenic patients. This smoking behavior is not seen in other mentally ill patients or in other people taking similar antipsychotic medications. The patients claim that smoking improves the clarity of their thoughts and their ability to pay attention to and remember the events of their lives.

2

WHY DO I FEEL THIS WAY?

"How do you feel?" You have asked, and been asked, this question many times. How does your brain answer this question? How you feel is determined by far more factors than your level of happiness or depression. Maybe you feel thirsty or hungry or cold; all of these would affect your answer to the question. Your answer to the question of how you feel is intimately connected with your survival.

The purpose of your emotions is to control evolutionarily conserved behaviors that are critical to your survival. If you are cold or hungry, you need to act upon this information in order to increase your likelihood of survival. For this reason, your brain has evolved a series of interwoven systems that work together with sensory inputs from inside your body to answer the question of how you feel; this brain network is called the limbic system. The limbic system controls many aspects of your survival, including the balance of energy and water, body temperature, hormones, sexual behavior, and your ability to experience pleasure. The limbic system also influences what you learn and remember. Your limbic system encourages the brain to remember those things, events, or people who pleased or frightened you in order to control future behaviors related to your survival. We rely upon our memory to make decisions about who we like, what foods made us sick, and what places or things frightened us.

A few cortical regions are considered to be part of the limbic system and play important roles in the expression of emotion; I will focus on just two. The first one is deep inside the brain and is called the cingulate gyrus. Imaging studies have discovered that this part of the brain determines the perceived level of pleasantness and unpleasantness of sensory stimuli, such as pain or the taste of chocolate. When you enjoy the smooth richness and flavor of a piece of chocolate, you can thank your cingulate gyrus. One Freudian psychologist summed up the function of the cingulate gyrus as where your superego and id compete to determine what you will do at any given moment. Another important role of the cingulate gyrus is the control of punishable behaviors; this region provides inhibitory control of behaviors that you have learned to avoid. For example, when you were young, you might have been punished for making loud sounds in public or jumping up and down on the sofa. Many years ago neurosurgeons discovered that if they destroyed small regions at the front end of the cingulate gyrus, patients were better able to control their symptoms of obsessive compulsive disorder, such as repeated handwashing.

The second cortical limbic region of interest is called the insula; this brain region interprets for us whether we like or dislike complex sensory inputs. The insula lies in a crevice, called the lateral fissure, deep on the side of the brain at about the level of the top of your ear. The insula is activated when we are listening to music that we like, when we hear the voice of someone we like, or when someone we like is stroking our arm. The insula also is activated by disgust, such as having a stranger on a bus begin stroking your arm or begin touching you, or by watching videos of unpleasant or repulsive images. I mentioned these two limbic cortical areas for another reason: the cingulate and insula are selectively activated when subjects report that their minds are wandering. Possibly, when our mind wanders, we activate these brain regions in order to judge the degree to which we like or dislike the contents of our daydream.

What is fear?

Judging what you like or dislike allows you to enjoy life. In addition, knowing what you should fear, and quickly recognizing the biological changes in your body that indicate fear, could save your life. This critical task is largely handled by a small almond-shaped structure, the amygdala, which lies deep within the bottom of the brain, not far from your ears. The amygdala receives information from many brain regions, your internal organs, and external sensory systems, such as your eyes and ears. The amygdala integrates this information with various internal drives, such as whether you are hungry or thirsty or in pain; it then assigns a level of emotional significance to whatever is going on. For example, when the amygdala becomes aware that you are alone and hearing unfamiliar sounds in the dark, it initiates a fear response, such as panic or anxiety. It then activates the appropriate body systems, the release of hormones, and specific behaviors to respond to the (real or imagined) threat. The amygdala also is activated by sensory stimuli that seem ambiguous or unfamiliar to us, such as unfamiliar sounds or people. In response to ambiguous or unfamiliar stimuli, we become vigilant and pay closer attention to what is happening in our immediate environment. If you were a dog, your ears would perk up. Your amygdala gathers as much sensory information as possible, compares it to what you already know, and then instructs other brain regions to respond.

Almost without fail, and regardless of the nature of the information gathered by your vigilant brain, the amygdala usually comes to the same conclusion: be afraid. If a sensory event, such as a sight or sound or taste, is unfamiliar; your limbic system almost always assumes that the situation is potentially dangerous and should be treated as such. If everything is assumed to be dangerous until proven otherwise, you are much more likely to survive the experience and pass on your be-fearful-first genes. Thus, humans fear everything that is unfamiliar or not-like-me: we fear unfamiliar dogs, people who look or dress differently, unfamiliar places, unfamiliar odors, things that go bump in the night, people who stare at us for too long, heights, enclosed small spaces, dark alleys, unknown people who follow us, etc. You get the idea. We all have witnessed the consequences of fear: we hide behind closed doors, we hide in protected or gated communities, we keep a loaded gun by every door and under the pillow, we hire bodyguards, we install security systems, we build walls. Brains evolved to perform one primary function: survival of the individual and the species; fear plays a critical role in survival. Unfortunately, your fear-inducing amygdala occasionally overreacts to trivial or harmless stimuli. Sometimes the amygdala induces behaviors that may get a person mentioned on the evening news.

Consider the following scenario: You are walking in an unfamiliar wooded area and you are aware of recent reports that snakes have been spotted along your current route. Then, without warning, you spot something brown, round, and coiled up on the ground next to a fallen tree. Your flight-or-fight response to this potential threat is activated immediately, quickly increasing your heart rate, respiration, and blood pressure; then, you realize that it is only a coil of discarded rope. Was your physiological response reasonable and appropriate? Yes, it was, because it prepared you to escape or defend yourself from a perceived danger. Your physiological response was so fast that it preceded recognition of the actual stimulus, the rope, due to the fact that your amygdala appears to receive partially processed sensory information before the more complex parts of your brain have had a chance to identify the true nature of the threat. Your brain evolved to help you survive to pass on your genes to the next generation. The best way to achieve this goal is to induce a response immediately to imagined threats regardless of whether that response is appropriate or not. Whether you are walking down a dark alley or are in a landscape full of snakes does not make any difference to your brain; you need to prepare yourself for fight or flight to defend your be-fearful-first genes so that you can pass those be-fearful-first genes along to your offspring.

By now you have clearly gotten the point that being frightened of everything all of the time is a safe and effective way to maintain your species. Unfortunately, it is also quite stressful, and chronic stress ultimately will have negative consequences upon your health. The brain, due to the impact of evolution, does not concern itself with the long-term effects of chronic stress on the body because these negative consequences usually appear long after you have finished reproducing and passing on your be-fearful-first genes to the next generation.

Due to its control over your emotional response, the amygdala plays a critical role in the decision-making processes in your brain. In order to achieve this goal, the amygdala influences the function of many other brain regions. It activates the frontal lobes of your brain to increase your vigilance to potential threats. The amygdala also controls how your brain processes sensory inputs that are associated with emotional experiences. This is an extremely important function because it determines whether you will remember the details of fearful events. For example, mugging victims tend to distort the details of the tragic event by "remembering" that the mugger was bigger and uglier, the gun was bigger, the alley was darker, etc. Recall my point from Chapter 1, the brain is not an accurate recording device; the influence of the amygdala makes memories more interesting or frightening than the events truly were. The influence of the amygdala, however, also makes it less likely that you will walk down that alley alone again. Your amygdala has succeeded again and your be-fearful-first genes live to breed another day!

The amygdala also becomes quite active when other people are looking at us. This response underlies why public speaking is usually rated as people's number-one fear; even more feared than heights, deep water, death, bugs, loneliness, and darkness. Neurons in the amygdala pay attention to the eyes of other people in order to inform you whether someone is staring at you. Staring at one's prey is a challenging action that is often a prelude to an attack. If someone in a crowded room, even if it was only a little girl holding her doll, started staring at you, how would that make you feel? She is following you, keeping her eyes trained on your every move; even her doll's eyes now seem focused on you. What does she want from you? Why is she following you? Feeling threatened yet? Yes, indeed! We all would respond with fear to a similar situation, no matter how innocent the "attacker" might seem.

Children with autism do not respond to staring; magnetic resonance imaging (MRI) scanning studies of their amygdalae indicate that these children primarily pay attention to the mouths of other people and, therefore, miss critical social cues. The amygdala in older people is less responsive to social cues, threatening or not, than it is in younger people. Why? Possibly because as we get older and experience numerous and varied emotional and fearful events, our frontal lobes gain more control over how the amygdala responds to incoming sensory information. Studies have shown that the frontal lobe is responsible for turning off the amygdala. Indeed, humans with a thicker frontal cortex (specifically, the bottom middle of the brain) have a greater ability to reduce activation of their amygdala that is prompted by strongly emotional stimuli. You might remember the television show *Star Trek* and the planet of Vulcans who were always in complete control of their emotions; possibly the Vulcans had big frontal lobes.

Studies of brains from preterm babies have shown that the amygdala is wired up to the rest of the brain prior to birth and is, therefore, capable of helping the brain store strong, usually fearful, emotional memories. This might explain why some adults have inexplicable fears. Possibly, at this very early stage of brain development, the amygdala was able to record a negative memory while the hippocampus was still unable to form a memory of the event associated with the unpleasant experience.

Why are close-talkers so frightening?

During the fifth season of the long-running and very popular sitcom *Seinfeld*, we were introduced to a person who was a "close-talker," a person who stands unusually close to others when speaking. We have all met this person. A colleague of mine used to stand so close when talking to me that his hand gestures actually occurred behind my head; they only could be appreciated by someone standing behind me! Is there an explanation for this behavior? Studies have found that bilateral damage to the amygdala reduces a person's need for interpersonal distance when talking to others. Although numerous social and cultural differences might explain why some people do not acknowledge your personal space when talking to you, some of these "close-talkers" might have a poorly functioning amygdala due to injury, mild hypoxia, or inheritance. Bilateral damage to the amygdala also may reduce a person's ability to produce a normal fear response to dangerous stimuli. In humans, one peculiar consequence of bilateral amygdala injury is the loss of the ability to laugh at other people's jokes; in contrast, these individuals are capable of laughing at their own jokes. Age-related changes in the amygdala thus might explain why your grandfather never laughs at your jokes; rest assured, however, you really are very funny.

What is depression?

Given the important role of the amygdala and hippocampus in the control of emotion, it is not surprising that scientists have discovered a connection between impaired function in the amygdala and hippocampus and symptoms of depression. Numerous studies have discovered that the hippocampus is smaller in patients with major depression as compared to those without the illness. The degree of brain shrinkage is influenced by how long the depression has been untreated. Imaging studies have shown that the activity of neurons within the amygdala also is significantly altered in patients with major depressive disorder. One day, noninvasive monitoring of the activity of these brain regions might offer better diagnostic accuracy for major depression and provide insight into whether specific treatments are truly helpful for this devastating brain disorder. A cure for depression is badly needed.

Each year, more than 100 million people worldwide develop clinically recognizable depression. The incidence of depression is 10 times greater than schizophrenia. Nearly twice as many women than men will report at least one major depressive episode during their lifetime. Depression is the leading cause of disability for women aged 15–44 years across all nations and cultures. The diagnosis of depression depends on three main factors: etiology, severity, and duration. Etiology refers to the cause of the depression. For example, depression may appear as a normal reaction to grief or as a primary affective disorder, as a reaction to the withdrawal of a drug, or as a component of a wide variety of medical problems, such as cancer or liver disease. Severity is a measure of how debilitating the symptoms are to the person's lifestyle. Everyone experiences occasional feelings of sadness that may last a few days; in contrast, major depression is characterized by long-term (usually longer than two weeks) feelings of hopelessness, irritability, and loss of interest or pleasure in most activities. In addition to these symptoms, depression often is associated with significant changes in appetite or weight, poor sleep quality, excessive feelings of guilt, feelings of worthlessness, lethargy, and recurrent thoughts of death or suicide attempts.

Over the past century, younger and younger people are being diagnosed with major depression. There are many possible reasons for this trend, such as increased life stressors, more sensitive diagnostic methods that are catching previously undiagnosed cases, and an increase in the number of patients, particularly males, willing to seek medical care at earlier ages. Depression has many causes; this has necessitated many different, and sometimes quite dangerous, treatments.

How is depression treated?

In 1786, the Italian physician Luigi Galvani recognized that our brains communicate by electrical signals; this led to the development of ways to utilize electricity to manipulate brain function. Electroconvulsive therapy (ECT) was first introduced in 1938 and involves applying a brief electrical pulse to the scalp while the patient is unconscious (we hope) under anesthesia. ECT was found effective for treating multiple psychiatric illnesses, especially depression. The experience of ECT sounds horrifying and the treatment still carries a highly negative stigma. Following the introduction of safer antidepressant medications, the use of ECT treatment declined during the 1960s. Its use has increased since the late 1970s because of improved delivery methods and increased safety and comfort measures. ECT most commonly is administered to patients who fail to respond to medications or who do not tolerate the side effects of standard medications. Most important, when patients demonstrate symptoms that increase the risk of harm to themselves or others, ECT may be the treatment of choice because its benefits are almost immediate.

In comparison, the non-ECT treatments currently available for the treatment of depression are only modestly effective, and treatment resistance remains a significant problem. Unfortunately, according to recent analyses, our current antidepressant medications are no more effective than those introduced over 50 years ago, although the newer drugs usually have fewer unpleasant or dangerous side effects. Because most patients show spontaneous recovery from their depression, especially when it is a reaction to life events, it is estimated that today's medications only provide for about 20%–30% more recoveries than if no drug were administered at all! Consider that statistic for a moment: Doing nothing works for almost 80% of people diagnosed with depression. The problem is that healthcare providers are not able to predict which patients fall into that 20%–30%; thus, today virtually all depressed patients are medicated.

Most drugs prescribed today to reduce the symptoms of depression (there is no cure) act by enhancing the action of the neurotransmitters serotonin, dopamine, or norepinephrine. These neurotransmitters are produced in your brain from the components of your diet. They are used by neurons to communicate with each other; serotonin, dopamine, and norepinephrine all play a role in determining how you feel. Current antidepressant medications block the inactivation of serotonin, norepinephrine, or dopamine within the synapse by blocking reuptake; thus, they are called selective reuptake inhibitors. If the drug selectively prevents the reuptake of serotonin, for example, it is given the acronym SSRI (selective serotonin reuptake inhibitor). Selective reuptake inhibitors slow the inactivation of neurotransmitters once they are released from neurons; this prolongs the time that they are available to act upon other neurons. Think of these drugs as door locks and the neurotransmitters as cats. You open the door and your cats run outside and interact with other cats in the neighborhood; you put an end to this playtime by opening the door and allowing the cats to rush back inside. Selective reuptake inhibitors keep the door locked so that the cats are forced to stay outside and play for a longer period of time. Ultimately, the neighborhood (your brain) is full of cats (e.g., serotonin) running around between the houses (neurons). However, simply because this is how these drugs *can* act within the brain does not offer any insight into how the drugs actually *do* reduce the symptoms of depression. Fortunately, the U.S. Food and Drug Administration (FDA) does not require that the therapeutic mechanism of a drug be understood, only that the drug works and is safe. The modern selective reuptake inhibitors are considered relatively safe and effective for most patients.

Why do you sleep so poorly when you are depressed?

Depression is often inherited and also can develop in association with other common mental or physical disorders. For example, depression, sleep disorders, and migraine headaches may all have similar underlying neural mechanisms; they often co-occur, more often in women than in men. In a recent study of patients with migraine, 58% reported also suffering anxiety regularly, 19% had chronic long-standing depression, and 83% had poor sleep quality with excessive daytime sleepiness. The connection between sleep and depression is fascinating because it might offer clues to understanding both phenomena. Let us take a look at one of these clues: nondepressed people start dreaming about two hours after falling asleep; depressed people start dreaming almost immediately. When depressed people start taking antidepressant medications, the time it takes for them to enter their first dream episode lengthens toward a normal duration. If this does not happen, the prognosis for recovery on that medication is poor. Sleep deprivation has a well-known antidepressant effect but only if one is depressed. As many of us learned during college, skipping a night of sleep will not make a nondepressed person happier. Indeed, it tends to make people anxious and irritable, reactions which are often symptoms of depression. The same applies to antidepressant medications; that is, if you are not depressed, taking an antidepressant drug will not make you happier. For many migraine sufferers, depression typically begins with the onset of their migraine and coincides with bouts of insomnia and anxiety. About 25% of patients with migraine have depression, and almost 50% of these patients also have anxiety. People with migraine, depression, and anxiety are more sensitive to subtle shifts in hormonal levels as well as to air travel, sleep loss, or specific foods; many of these act as triggers for one or more of these symptoms. The genes that link these various disorders have not been discovered.

Why is depression so common?

Depression has been called the common cold of mental illnesses because all of us catch it at some point in our lives. This is a fitting metaphor given what is now known about one potential cause of depression. An imbalance in gut bacteria and viruses may underlie depression. Therefore, depression may be so common because an imbalance in the variety of gut bacteria is so easy to induce. Lifestyle choices, such as shift work or smoking, the overuse of antibiotics, or disease conditions such as irritable bowel syndrome, are associated with depression and a reduction in the variety of gut bacteria.

Recent investigations of humans and locusts have focused upon the effects of bacterial infection, the development of body and brain inflammation, and the appearance of sickness behavior or depression. Recent studies show that the level of proinflammatory proteins in the blood increase during depression. Most of these proinflammatory proteins can cross the blood–brain barrier easily and influence brain function. In addition, the current antidepressant medications exhibit anti-inflammatory proclivities. Moreover, obese humans produce significant amounts of proinflammatory proteins. Taken together, these findings may explain why obesity and depression occur so often in advanced industrial economies and often are inherited together. Furthermore, obese humans do not respond well to most antidepressant drugs, as compared to nonobese depressed patients. More on this later.

What is the role of serotonin in depression?

Humans and locusts use the neurotransmitter serotonin for apparently similar functions. When serotonin levels are too low due to the content of their diet or stress from overcrowding, both locusts and humans display solitary behaviors and make an effort to become isolated from others. When humans consume diets that are low in tryptophan, a condition often seen when someone first goes on a vegetarian or vegan diet, the brain produces much less serotonin and humans display many of the symptoms of depression such as anxiety, irritability, and difficulty thinking. When locusts eat grain containing tryptophan, their brains make more serotonin and they become gregarious and spend time in the company of large crowds of other locusts aggressively eating tryptophan-containing fields of crops. When depressed humans take medications that enhance serotonin function in their brains, they become happier and enjoy the company of friends and family. Depression and sickness behavior (malaise, failure to concentrate, sleepiness, fatigue, coldness, muscle and joint aches, and reduced appetite) share many features that are regulated by infection status, level of inflammatory proteins being produced, and availability of specific nutrients in the diet that lead to the production of serotonin. Sickness behaviors are initiated by our immune system in response to infection or injury; their goal is to remove us from potential harm while we heal.

Taken together, these similarities suggest that depressive behaviors are so common because they evolved as a universally needed survival mechanism that was passed on from generation to generation as an effective method of surviving in a dangerous world teeming with opportunities for injury and infection. All species, from locusts to humans, may have evolved the ability to suffer depression because it has survival value for the individual, and, therefore, the species. However, due to the negative health consequences of long-term inflammation in the brain and body, the symptoms of depression must be aggressively treated because untreated depression increases the risk of cancer and reduced lifespan. Ironically, depression-related behaviors may be evolved physiological reactions that are healthful in the short term but harmful if allowed to continue for a long time.

What is bipolar disorder?

Patients with bipolar disorder show many of the same symptoms of depression as described earlier. In addition to the symptoms of major depression, in order to be diagnosed with bipolar disorder a person must display at least one episode of mania. Mania is characterized by inflated self-esteem or grandiosity, decreased need for sleep, pressure to keep talking, racing thoughts, easy distractibility, excessive irritability, agitation, and attention to pleasurable activities that often have a high probability for negative consequences, such as unrestrained buying sprees, sexual indiscretions, or foolish investments. Many artists, writers, poets, and composers, such as Virginia Woolf, Jack London, Jackson Pollock, and Ernest Hemingway, who suffered with bipolar disorder produced some of their most iconic works during manic episodes.

The initial symptoms of bipolar illness usually begin in late adolescence, often presenting as depression during the teen years or even earlier. Children with bipolar parents have a significantly increased risk of developing the illness. Too often, the symptoms in children are difficult to recognize because they can be mistaken for normal age-appropriate behaviors. No single genetic marker has been identified; however, recent studies suggest some genetic similarities among bipolar illness, schizophrenia, and autism. Bipolar disorder is a collection of symptoms that represent a disruption in how multiple brain regions communicate with one another. An error in how some brain regions wired themselves together during development makes some people susceptible to bipolar disorder. Possibly as a consequence of these wiring problems, some brain regions are too small and some areas of cortex are too thin.

Noninvasive investigations of the brains of patients with bipolar illness have shown that a portion of the frontal lobe is reduced in size. Preliminary studies suggest that electrical stimulation of this part of the brain can compensate for the atrophy and reduce the symptoms of depression in bipolar patients who are otherwise treatment resistant. The drug treatments for patients with bipolar disorder are known collectively as mood-stabilizing drugs and include lithium salts and the anticonvulsants valproate, carbamazepine, and lamotrigine. For more than five decades the primary treatment for bipolar disorder has been lithium salts. Lithium increases the time between manic episodes; it has minimal effects upon the depression phase of this illness. However, by reducing the incidence of mania, the incidence of depression also is reduced. This discovery suggests that the cycling between mania and depression involves neural mechanisms that are somehow interconnected with each other in the brain. Lithium has many known actions within the brain; whether any of these is responsible for its antimanic proclivities is unknown. In general, lithium treatment alone is insufficient to treat all of the symptoms of the disorder; the majority of patients are given a combination therapy of lithium and other medications such as antidepressants. Unfortunately, many patients cannot tolerate the unpleasant side effects of lithium therapy, including changes in weight and appetite, tremor, blurred vision, metallic tastes, and dizziness. Obviously, treatments that are more effective, act faster, and are tolerated better are needed badly. Newer classes of drugs that modulate the action of the neurotransmitter glutamate are now being actively investigated. In addition, a recent report claimed that deep brain stimulation, similar to that used on patients with Parkinson's disease, might be quite effective for patients with bipolar disorder.

Bipolar disorder currently is believed to arise from interactions between the two forces that drive most of our biology: the genetic risk factors we inherited from our parents and the consequences of our environment. It is now clear that unpleasant events during childhood that lead to chronic stress or mental and physical trauma also contribute to the appearance of symptoms later in life. Studies using both MRI scans and postmortem investigations indicate that if adverse experiences occur during critical developmental periods, actual structural and functional changes develop in the brain. These changes may have long-lasting effects on adult brain function. Furthermore, developmental errors in brain wiring can be aggravated or unmasked later in life by exposure to stressful events. Clearly, exposure to significant stressors is a key risk factor in the appearance of bipolar symptoms.

The cause of bipolar disorder is unknown, although it has a clear genetic component given that it tends to run in families. Most patients are diagnosed in their early twenties; however, for women the initial symptoms of the disorder may not appear until later in life around the onset of menopause. Overall, lifestyle factors likely interact with a complex blend of hormonal changes and genetic mutations. Imaging and genetic studies have identified interesting similarities between bipolar disorder and schizophrenia that might shed additional light on this complex disorder of mood.

What is schizophrenia?

The onset of numerous mental disorders peaks between the time of late adolescence and young adulthood, including attention deficit hyperactivity disorder, anxiety and mood disorders, schizophrenia, and substance abuse. What is the brain doing during this phase of life that coincides with the onset of so many disorders of higher mental function? One possible contributor is the completion of myelination of the frontal lobes that occurs during our mid-twenties. Myelination is analogous to the insulation on the wiring in your house; if it is not present, the wiring does not work correctly, or at all. Once a brain region finishes its myelination process, imaging studies have shown that the region becomes more active. Apparently, the complex interplay of neurons in your brain works best when their connections, that is, their axons, are fully insulated with myelin. The frontal lobes are the last region of the brain to finish this process of myelination; women finish by age 25 while men finish this process by age 30. You can see what this implies: on average, women have functioning frontal lobes at least five years sooner than men do. In general, men are diagnosed with schizophrenia earlier than women; however, the incidence of schizophrenia is higher in women after age 30. Scientists speculate that the problems in brain function become apparent, and thus more likely to be diagnosed, as the affected parts of the brain become fully active with development and maturation.

The latest studies of susceptibility genes suggest that attention deficit disorders, anxiety, depression, bipolar illness, and schizophrenia share some key genetic components related to neural development. By age 18, about 20% of adolescents will show symptoms of a mental illness that will persist into adulthood. For example, oppositional defiant disorder, which tends to occur in families with a history of attention deficit or mood disorders, usually appears during early childhood; approximately 90% of these children will develop schizophrenia as adults. Human genome studies indicate that schizophrenia has a strong genetic component that may involve the function of hundreds of unique genes related to development and neuroplasticity. Many different environmental influences also have been identified as risk factors. Recent theories of schizophrenia invoke a dysregulation of dopamine and glutamate neural systems, particularly within the frontal lobes. This dysregulation leads to a failure of the frontal cortex to control limbic function and may underlie the characteristic cognitive symptoms of schizophrenia. Noninvasive techniques have allowed scientists to look for changes in glutamate function within vulnerable brain regions and then make fairly accurate predictions regarding which patients will undergo significant remission of symptoms, as well as which are not likely to show significant remission. This information allows psychiatrists to make informed judgments regarding patient therapy.

Why do schizophrenics hear voices?

Here is part of a long e-mail message from a schizophrenic patient who displays many common aspects of the disorder, in particular hearing voices.

> Dr. Wenk – I'm not sure if you'd have any clue about this, but if you do, please contact me. I believe I was injected with some sedating drug while sleeping in my bedroom, having a vague memory of partially waking up to the event. A few months later I began hearing soft voices that could communicate with me, and I figured that someone had implanted a remote radio brain probe in my head. In July I began to be extensively harassed, violated, and at times tortured via this device.

Why do schizophrenics hear voices? Why do antipsychotic medications that block dopamine receptors usually alleviate this symptom? A recent study offered some potential answers to both questions. The study identified a significant disruption of neural circuitry within the auditory (sound processing) cortex in an animal model of schizophrenia. These neurons of the auditory cortex become inappropriately active when schizophrenic patients are hallucinating voices talking to them.

Schizophrenics may hear either hostile voices goading them to jump off a bridge or a mother's soothing words of advice; which type they hear depends on the cultures in which they live. In the United States, schizophrenics report hallucinations of disembodied voices that hurl insults and make violent commands. In India and Ghana, however, schizophrenics report quite positive relationships with hallucinated voices that they recognize as those of family members or God. Interestingly, schizophrenia tends to be more severe and long lasting in the United States than in India.

The gene that underlies the presence of auditory hallucinations may be responsible for the production of a specific dopamine receptor. Schizophrenics appear to have too many of these dopamine receptors; therefore, it is not surprising that medications that selectively block dopamine receptors can reduce the frequency of these auditory hallucinations effectively. This is just a single example of how systematic studies of brain chemistry and physiology are slowly advancing scientists' understanding of the complex interactions of the genetic, neurochemical, and anatomical changes that underlie this disorder. Hopefully, the results of these investigations will lead to better treatments for schizophrenic patients.

Are dolphins schizophrenic?

Scientists have wondered whether the occurrence of schizophrenia and autism in humans is due to the rapid evolution of our brains. Are psychiatric diseases the cost of the higher brain function in humans? A study of the molecular evolution of the genes associated with schizophrenia, autism, and other neuropsychiatric diseases compared across mammalian species and among disease classes, with a focus on primate (chimpanzee, bonobo, gorilla, orangutan, gibbon, macaque, baboon, marmoset, and squirrel monkey) and human lineages. This study concluded that genes associated with schizophrenia and autism are not evolving uniquely or more frequently in humans. Interestingly, one species stood out in the analysis as possessing a much higher number of both schizophrenia and autism genes: the bottlenose dolphin. Can a dolphin experience paranoia? No one knows. It is worth considering, however, that the surprisingly intelligent behaviors of these mammals are due to the presence of some unexpected genes.

How is schizophrenia treated?

Whatever the causes of schizophrenia might be, almost universally, the treatment is to block dopamine receptors. Does this mean that schizophrenia is due to a problem with dopamine? No, not at all. In fact, an alteration in dopamine function probably does not cause schizophrenia; rather, the symptoms are most likely just a secondary consequence of alterations of some other neural system, such as glutamate, in the brain. This may explain why the blockade of some dopamine receptors within the brain reduces the severity of a few bothersome symptoms, but not others. The antagonism of dopamine receptors simply may compensate for the presence of an error of chemistry that exists somewhere in the brain. All neuroscientists know for certain is that whatever the reason may be for their efficacy, antipsychotics that block dopamine receptors provide significant benefits for some, but not all, schizophrenic patients. Unfortunately, these drugs—especially the antipsychotics introduced in the 1950s—have side effects that make these patients move as though they have Parkinson's disease. Given the very unpleasant side effects of these drugs, it is easy to appreciate why so many schizophrenics hate taking their medications. The side effects of dopamine receptor blockade occur rather quickly, but the clinical benefits require two to three weeks, or longer, to develop fully. The time it takes for these drugs to produce a noticeable benefit implies that compensatory changes in brain function are required for these drugs to produce clinical benefits in schizophrenic patients. The nature of these compensatory changes is not understood.

3

HOW DO FOOD AND DRUGS INFLUENCE MY BRAIN?

The brain is the organ of your mind; therefore, food and drugs can have a profound influence on how you think, act, and feel. These effects can be profound, subtle, or barely noticeable. Why do some chemicals in your diet affect your brain and how you feel, while others do not? Many drugs or nutrients that potentially might influence brain function are never able to enter the brain because of the presence of a series of barriers; the most important of these is the blood–brain barrier.

This barrier allows the easy entry of drugs and nutrients that are lipid-soluble (i.e., fat-soluble) and restricts the entry of drugs and nutrients that are water-soluble. Extremely lipid-soluble drugs enter the brain rapidly; they also tend to exit rather rapidly, which reduces the duration of their action. Some familiar examples of lipid-soluble nutrients are the vitamins A, D, E, and K. Nicotine and caffeine are also quite lipid-soluble and enter the brain easily; if they did not, then it is highly unlikely that anyone would bother consuming them so often. Take a moment to appreciate how this fact has been an incredible boon to the evolutionary success of tobacco and coffee plants: their discovery by our species, coupled with the fortuitous nature of our brain chemistry, led to their widespread cultivation and protection as two of the most important plants on earth. Human behavior has impacted these plants as much as they have impacted human history; for example, the introduction of coffee and tea fueled the Enlightenment and the Industrial Revolution.

This chapter will discuss the foods and drugs that affect your brain and, thereby, your behavior. The distinction between what is considered a drug (i.e., something that your body wants or needs to function optimally) and food (i.e., something that your body wants or needs to function optimally) is becoming increasingly difficult to define. Indeed, the routine use of some substances, such as stimulants and depressants, is so universal that most of us do not even consider them to be drugs but, rather, actual food. Are coffee, tea, tobacco, alcohol, cocoa, or marijuana nutrients or drugs? For many people, the distinction has become rather meaningless because their body craves many of these substances at all times. Obviously, anything you take into your body should be considered a drug whether it is nutritious or not. For the remainder of this chapter, I will make no distinction between drugs and food: they are essentially just chemicals that have unique effects on the body.

The foods we eat and many of our most popular psychoactive drugs often come from plants. Many plants contain chemicals that are very similar to the chemicals in our brains. The similar nature of these chemicals underlies why the contents of our diets can influence brain function.

Why do plants affect the human brain?

Plants produce chemicals that are capable of affecting our brain because they share an evolutionary history with us on this planet. Even primitive one-celled organisms produce many of the same chemicals that are in your brain. Therefore, whether you choose to eat a bunch of broccoli or a large pile of amoeba, the chemicals they contain may alter how your neurons function and, therefore, how you feel or think.

The fact that you share an evolutionary history with insects and reptiles also underlies the ability of venoms to produce the unpleasant affects you feel if you are stung by a bee or bitten by a snake. The bugs add serotonin to their venom in order to increase blood flow to the site of the bite or sting, thus increasing the chances that you will absorb most of the venom. Our shared history with plants and animals here on earth leads to some interesting predictions. For example, consider the following science fiction scenario: A spaceman is walking on an earth-like planet and is suddenly bitten by an unfriendly and grizzly looking creature. The spaceman can see that he is injured and that a liquid substance was injected under his skin by the beast. Does he die? No, he does not die, because his species and that of the creature on this foreign planet do not share an evolutionary past or a common ancestor. Their independent evolutionary paths make it highly improbable that they use similar neurotransmitter molecules within their respective brains and bodies.

Back on earth, people in ancient cultures were certainly very aware of the unique properties of certain plants and of the consequences of consuming them; indeed, they often sought them out as remedies for a variety of physical illnesses. This ancient use of plant extracts as medicines was also likely the beginning of a long series of reforms in our concept of how the brain functions and what its role is as the organ of the mind. For example, the realization that it might be possible to treat mental illness in the same way that one treats physical illness—that is, by using drugs or diet—was slow to gain general approval in part because of the wide-ranging, and for some still quite frightening, implications about what this meant regarding the nature of the human mind. Our grandchildren will likely have a whole host of highly modified chemicals added to their diets strategically designed to enhance a broad range of mental functions. In fact, we already do have a vast pharmacopeia, legal and otherwise, that can affect the brain, and no end of debate about its value and effectiveness.

Three basic principles apply to any substance you ingest that might affect your brain. First, these substances should not be viewed as being either "good" or "bad." Drugs and nutrients in your diet are simply chemicals—no more, no less. They initiate actions within your brain that you either desire or would like to avoid. Second, everything you consume likely has multiple effects. Because your brain and body are so complex and because the chemicals you ingest are free to act in many different areas of your brain and body at the same time, they will often have many different effects—both direct and indirect—on your brain function and behavior. Third, the effect of a drug or nutrient on your brain always depends on the amount consumed. Varying the dose of any particular chemical changes the magnitude and the character of its effects. This principle is called the "dose–response effect"; that is, in general, greater doses lead to greater effects on your brain. Sometimes, however, greater doses produce completely opposite effects from those of lower doses. For example, aspirin reduces body temperature when taken at normal therapeutic doses but increases body temperature when taken at high doses.

How do we become addicted to specific foods and drugs?

Sometimes the effects of certain chemicals are present in the brain for so long that the brain slowly adjusts to their presence. Over time, the brain acts as though the drug or nutrient has become a necessary component of normal brain function. You experience your brain's adjustment to the eventual absence of this substance as craving. Consider, for example, the very powerful drug, sugar. Your brain needs sugar (usually in the form of glucose) to function normally. The many billions of neurons in your brain require a constant supply of glucose to maintain their ability to produce energy and communicate with other neurons. The brain consumes the equivalent of about 12 donuts worth of glucose every day. Neurons can tolerate a deprivation of glucose for only a few minutes before they begin to die. Therefore, as blood levels of sugar decrease with the passage of time since your last meal, you begin to experience a craving for food, preferably something sweet. The presence of sugar in your brain is considered normal, and its absence leads to the feeling of craving and the initiation of hunting or foraging behaviors, such as seeking out a vending machine for a chocolate bar. If you wish to experience the truly overwhelming and powerful nature of drug craving, just stop eating for a full day.

Now you can easily recognize a parallel with the experience of a heroin addict. Within a few hours you will not be able to think of anything but food (heroin), you will do anything, sell anything, or steal from anyone to get food (heroin); as time passes, nothing is more important to you than the next meal (shot of heroin). The brain behaves as though it cannot tell the difference between a food and a drug; they are both just chemicals. The constant consumption of caffeine, nicotine, or almost any chemical can produce similar types of compensatory changes within your brain and lead to craving when they are absent from the brain. This response is exactly what your brain evolved to do for you: to be flexible and learn how to survive, to adapt to a changing environment, and to adapt to the variety of chemicals that you consume. When this situation of "normalcy" is lost because of the absence of something that your brain has become accustomed to having regularly available (e.g., sugar, amphetamine, heroin, or anything else that you are accustomed to consuming), your brain reacts by creating in you the urge to replenish its supply. You experience this feeling as a craving, regardless of the legality, safety, or cost of the substance being craved.

Can I blame it on my parents?

The effects of a drug or your diet on your brain are greatly influenced by your genes, the nature of the drug-taking experience, and the expectations you have about the consequences of the experience. For example, if you respond strongly to one drug, you are likely to respond strongly to many drugs, and this trait is likely shared by at least one of your parents. Many biological factors such as age and weight play a crucial role in the way that food and drugs affect the brain and influence behavior. So, too, does the unique neural circuitry that you inherited from your parents and that neural circuitry sometimes influences whether a drug will be exciting or depressing to you.

This concept is known as the *Law of Initial Value*. According to this law, each person has an initial level of excitation that is determined by his or her genetics, physiology, sickness or health status, drug history, and environmental factors; the degree of response to a psychoactive drug depends on how all of these factors affect one's current level of excitation. For example, patients suffering from pain, anxiety, or tension experience euphoria when they are given small doses of morphine. In contrast, a similar dose of morphine given to a happy, pain-free individual often precipitates mild anxiety and fear. If you have a fever, aspirin lowers your body temperature, but aspirin cannot cool your body on a hot day—you must first have the fever for it to work. Coffee produces elation and improves your ability to pay attention if you have been awake for a long period of time or had poor sleep the night before; in contrast, the same dose of coffee is likely to produce much less arousal, or even anxiety, if you are well rested. Catatonic patients may respond with a burst of animation and spontaneity to an intravenous injection of barbiturates, whereas most people would simply fall asleep. Sedative drugs create more anxiety in outgoing, athletic people than they do in introverted intellectual types. Obviously, it can be quite difficult to predict how some drugs or nutrients will affect your brain function. It is not safe to rely upon the experience of others; their physiology and genetic history are likely very different from yours. This principle of brain function is true even for very commonly used, apparently safe drugs, such as caffeine.

Why am I addicted to caffeine?

Studies have documented that caffeine consumption in young adults directly correlated with increased illicit drug use and generally riskier behaviors; however, these correlational studies never examined the long-term consequences of caffeine consumption. For example, does long-term coffee consumption during adolescence lead to riskier behaviors during adulthood? How might caffeine consumption produce such long-lasting changes? The answers lie in understanding the actions of caffeine in the brain. In adults, caffeine appears to enhance indirectly the activity of dopamine within the brain's pleasure centers. Drinking coffee produces a mild euphoria due to this effect and encourages the brain to crave more coffee. Yes, coffee is addictive, but only mildly so as compared to many other drugs of abuse such as tobacco and cocaine.

The adolescent brain responds differently to caffeine as compared with the adult brain. For example, caffeine produces a more dramatic increase in motor activity in adolescents. In addition, long-term caffeine consumption produces more tolerance faster in adolescents as compared with adults. This suggests that caffeine might produce greater changes in brain chemistry in the developing adolescent brain. This speculation was strengthened by the finding that long-term caffeine consumption during adolescence leads to greater sensitivity to amphetamine-like drugs that are used to treat attention deficit hyperactivity disorders. Fortunately, there is no current evidence that caffeine consumption leads to attention deficit hyperactivity disorders in children.

A recent study determined that long-term caffeine consumption during adolescence altered the brain's neurochemistry so that in adulthood the brain's response to cocaine was enhanced. In contrast, consuming caffeine as an adult does not produce the same type of enhanced response to cocaine. This finding suggests that the developing adolescent brain is vulnerable to the effects of caffeine and that these changes can linger into adulthood and increase the abuse potential of euphoria-producing drugs such as cocaine. Therefore, by definition, coffee and tea are gateway drugs to cocaine. Thus, is caffeine a drug or a food? Sometimes it is very hard to tell the difference.

What should I eat to feel better?

Almost everything you consume may directly or indirectly affect brain function. In order to understand better how food and drugs affect the brain, it will be helpful to divide them into three categories. First, there are chemicals we consume that produce almost immediate effects on brain function, such as those found in coffee, heroin, alcohol, nicotine, marijuana, some spices, and a few psychoactive plants and mushrooms. Their effects depend on how much of the chemical reaches the brain. Sometimes the amount of chemical that actually enters the brain is so low that we do not notice its effects. For example, consider nutmeg: Low doses will be in pies next Thanksgiving, and most of us will not notice that it contains a chemical that our bodies convert into the popular street drug Ecstasy. Yet, if you consume the entire canister of the spice, your intestines will notice (with a terrible diarrhea), and there is a good chance, if the nutmeg is fresh, that you will hallucinate for the next 48 hours. Nutmeg abuse is often popular in prisons.

Second, there are those foods that affect our brain slowly over a period of a few days to many weeks. This is usually called "precursor-loading" and would include many different amino acids (tryptophan and lysine are good examples); carbohydrates that have a high glycemic index such as potatoes, bagels, and rice; fava beans; some minerals (iron and magnesium, in particular); lecithin-containing products such as donuts, eggs, and cakes; chocolate; and the water-soluble vitamins. The purpose of these foods is to bias the function of a specific neurotransmitter system, usually to enhance its function in the brain. For example, scientists once thought that drinking a glass of warm milk before bed or eating a large meal of protein made us drowsy because of tryptophan loading. The current evidence does not support this explanation, but the claim makes my major point: You must get enough of any particular nutrient or chemical to the right place, and at the right dose, in your brain in order for you to notice any effects. In fact, tryptophan has difficulty getting into your brain, particularly when consumed within the context of a large variety of other amino acids, as are present in meat, such as turkey, which, in fact, does not contain very high levels of tryptophan. Pumpkin seeds and egg whites contain far more tryptophan than turkey and no one has claimed that these food sources cause drowsiness. So, what is the scientific evidence for considering the cognitive effects of these foods? Mostly, it is related to what happens when we do not get enough of them. For example, studies have shown that consuming too little tryptophan makes us depressed and angry; historians now blame low-tryptophan diets for multiple wars and acts of cannibalism. Too little of water-soluble vitamins (the B's and C) in the diet will induce changes in brain function that we will begin to notice after a few weeks of deprivation. Ordinarily, the foods in this second category require more time to affect our brains than do foods in the first category.

The third category includes the slow-acting, lifetime dosing nutrients. This category includes the antioxidant-rich foods such as colorful fruit and vegetables, fish and olive oils, fruit juices, anti-inflammatory plants and drugs such as aspirin, some steroids, cinnamon and some other spices, nicotine, caffeine and chocolate, the fat-soluble vitamins, nuts, legumes, beer, and red wine. People who eat these foods benefit from consuming them regularly over their life span.

The benefit comes from the fact that all of these foods provide our brains with some form of protection against the most deadly thing we expose ourselves to every day—namely, oxygen. Because we consume food, we must consume oxygen.

Because we consume oxygen, our tissues suffer the consequences. Thus, people who live the longest tend to eat foods rich in antioxidants or simply eat much less food. Although nicotine and caffeine prevent the toxic actions of oxygen in our brain, this should not be taken as a recommendation to smoke a cigarette with your morning coffee.

We can see here that depending upon how we frame the question about how food affects the brain we end up with a different list of foods and a different reason for consuming them. If you wish to alter your current brain function or slow your brain's aging, you need to eat specific foods. In truth, no one ever considers these distinctions when eating—most people simply eat what tastes good, and our brains evolved to reward us for eating sugar, fat, and salt. Consequently, food, like any illicit or licit drug, has both negative and positive effects; it all depends on what drug or food you consume and how much you consume.

How do I stop eating so much food?

The real challenge for your brain is how to stop you from eating. This decision is partly determined by how fat you are. The brain learns about this through the action of two hormones—leptin and insulin—and responds by reducing food consumption. The blood levels of insulin and leptin are continuously elevated in the brains of many obese people, but their brains ignore these hormonal signals and so eating continues. The effectiveness of these hormones is influenced by fluctuating levels of estrogen; this leads to the gender dichotomy that females are more sensitive to the appetite-suppressant action of leptin (initiated by their body fat), whereas males are more sensitive to the appetite-suppressant action of insulin (induced by eating). When it comes to food, female brains do not follow the same rules as male brains.

Your brain also gets sensory feedback from your mouth and nose about the smell, taste, and feel of the food, as well as the expansion of the stomach. Unfortunately, these signals can easily be ignored by the brain—and so we keep eating. ۱۱New research on how the brain gets us to stop eating has led to the development of drugs designed to reduce food intake by mimicking one or more of these feedback signals. But each time the same thing happens—caloric intake decreases for a short time and then the brain adapts to ignore the false signal; ultimately, regular caloric intake is restored. Why? Because the consequences of not ingesting a sufficient number of calories has terrible consequences for your survival. There is no evolutionary advantage to trying to lose weight by restricting eating. Four billion years of evolution have led to the following simple directive for all living things: Find and consume the energy within food, repeat often.۱۱

When an energy source is on the tongue, the brain is informed via a series of simple molecular interactions within the taste buds, which lead to the activation of reward pathways in the brain that utilize the neurotransmitters dopamine, endorphins, endocannabinoids, and orexin. Orexin was discovered only recently; it influences both our level of arousal and our craving for food. Take a moment to appreciate how orexin optimizes your daily existence and survival. Orexin-releasing neurons wake you in the morning and then make you crave food. Once food reaches your gut, it encounters still more receptors that detect sweetness, fattiness, and bitterness. It appears as though your entire gut is a continuation of the tongue with specialized taste receptors. The activation of these receptors slows the intestinal transit of the food, providing a greater opportunity for nutrient extraction within the limited length of the intestines.

Is there a good time of day to eat?

What would happen if you could only eat between the hours of 9 A.M. and 4 P.M.? Would you gain less weight and be healthier overall even if you ate a high-fat diet? The answer is yes and is based on how your body is influenced by your daily rhythms of eating and sleeping. There are always negative consequences to ignoring the role of your biorhythms. Many studies have documented that nightshift work, and the odd patterns of sleeping and waking that this lifestyle involves, has many negative health consequences, including insomnia, high blood pressure, obesity, high triglyceride levels, and diabetes—collectively known as the metabolic syndrome. In a recent study, mice were given free access to a nutritionally balanced diet or a diet that was high (61% of their daily calories!) in fat. Some mice were allowed total access to the food at all times; others were only allowed access for an eight-hour window during the early phase of their normal active period. Mice given all-day access to a high-fat diet (which the authors termed the standard American diet) developed obesity, diabetes, metabolic syndrome, and poor sleep-wake rhythms. Now for the good news! The mice that had time-restricted access to the high-fat diet were significantly healthier than the mice given all-day access to the same diet. These lucky mice lost body fat and had normal glucose tolerance, reduced serum cholesterol, improved motor function, and normal sleep cycles. Most surprising, the daily total caloric intake of all groups did not differ, regardless of their diet or feeding schedule.

Therefore, it truly does matter when you eat. The take-home message is eat early, skip dinner, and never have late-night snacks. Skipping breakfast and then overeating in the evening play a significant role in weight gain and obesity. Furthermore, people who skip breakfast report not feeling as satisfied by their food and being hungry between meals. If this sounds like you, then it's time to change your mealtimes.

What about carbohydrates?

A carbohydrate is a molecule made of carbon, hydrogen, and oxygen. Glucose is a carbohydrate and is commonly called "sugar." The adult brain has a very high energy demand requiring continuous delivery of glucose from the bloodstream. The brain accounts for approximately 2% of our body weight but consumes approximately 20% of glucose-derived energy, making it the main consumer of glucose. The largest proportion of energy in the brain is consumed by your neurons when they are busy processing incoming sensory information, thinking about complex problems, or contemplating your future. Your brain needs a constant supply of sugar; without it you would quickly lose the ability to think and slip into a coma. We must obtain this sugar from our diet. Somewhere in our evolutionary history, we lost the ability to convert fat into sugar; unlike a few fortunate animals, humans cannot perform this metabolic trick. So, in the morning when you wake up from a long period of fasting, your brain wants you to eat lots of sugar and other simple carbohydrate sources, such as a donut. There is a reason that donuts and sugar-laden cereals are so popular, and you can lay the blame on neurons within the feeding center of your hypothalamus. This mechanism works nicely. First thing in the morning, you eat lots of simple, easily digestible sugars and your brain rewards you with a good feeling by releasing dopamine and endogenous opiates. The amount of dopamine released into your reward center is proportional to how hungry you are; that might explain why you enjoy sugary cereals or a donut for breakfast—they simply taste much better after you have been fasting all night. Your brain needs the sugar to produce chemicals that are critical for learning and memory.

Impaired glucose regulation correlates with impaired learning in the elderly and is associated with Alzheimer's disease. Scientists recently have discovered that the inability of specific brain regions to use glucose efficiently precedes the degeneration of those same brain region decades later. Eating more sugar is not the answer to preventing dementia. Indeed, consuming large amounts of sugar is not healthy for your pancreas or cardiovascular system. What's good for the brain is not always good for the other organs of your body.

What about fats?

We have a fatty brain and fat plays many vital roles in brain function. In the past, very little attention was given to the influence of dietary fats on our mental state. Recent evidence indicates that it might be possible to manipulate our dietary fat intake to treat or prevent disorders of cognitive function. A recent study compared the effects of monounsaturated fats from olive and canola oils with polyunsaturated fats from meat, fish, and vegetable oils on a variety of biochemical changes and electrical properties of cells within a brain region that is critical for learning and memory. After 11 months, a diet high in monounsaturated fats, often referred to as the Mediterranean diet, altered brain chemistry in such a way that learning was enhanced, age-related cognitive decline slowed, and the risk of getting Alzheimer's disease was reduced. These findings support the addition of canola, olive, and fish oils to our diet and further demonstrate that sensible nutritional choices are vital for optimal brain function and good mental health.

Omega-3 fatty acids are a family of fats that are important components of the human diet. Some recent studies have concluded that being deficient in omega-3 fatty acids may affect brain physiology and increase the risk of cognitive decline. Superficially, this claim makes sense. After all, omega-3 is abundant in the brain and is involved in numerous critical functions. It also may enhance learning and memory processes in the brain. It has been argued that dietary intake of omega-3s, mainly from fish, can slow cognitive decline and the incidence of dementia. These claims may or may not be true. The problem is that the clinical trials related to these claims have either included too few patients or were conducted for quite brief periods of time. Thus, the results were highly variable and potentially misleading. Recently, a study investigating the potential benefit of omega-3s followed almost 3,000 people, aged 60 to 80 years, for 40 months. Their daily diets, medications, and health status were carefully monitored. The patients and their controls were carefully matched for education level, smoking habits, and alcohol use, among other features. The results confirmed that prolonged omega-3 intake (as fish or pill supplement) provides no significant health benefits. Cognitive decline also was unaffected. What does this mean? That a single, good dietary habit, such as high levels of specific essential nutrients, is not enough to provide protection for your aging brain.

In contrast to their lack of benefit for age-related cognitive decline, omega-3 fatty acids may have a beneficial influence on the outcome of depressive disorders. Chronic dietary supplementation with omega-3 fatty acids has produced antidepressant-like effects similar to those of common antidepressant drugs. The therapeutic approach of combining omega-3 fatty acids with low doses of antidepressants might lead to benefits in the treatment of depression, especially among patients with depression resistant to conventional treatments. Such an approach also could decrease the magnitude of some antidepressant dose-dependent side effects.

How does obesity affect brain function and development?

Scientists have demonstrated that obesity leads to hypertension, diabetes, sleep apnea, and numerous arthritic disorders. Obese individuals also perform worse in cognitive tests even when controlled for education level and evidence of depression. Furthermore, women who eat an unhealthy high-fat diet prior to and during pregnancy are more likely to give birth to children, particularly males, who are at risk of abnormal behaviors, predominantly anxiety, during adulthood. Physicians frequently warn pregnant women to monitor their caloric intake and maintain a healthy weight before and during pregnancy. Maternal nutritional status, infection, and physical or psychological trauma during pregnancy can all increase the risk of obesity, diabetes, and mental disorders in offspring. In the past, the concern was maternal malnutrition—that is, the developing fetus might lack critical nutrients for normal growth. Today, in the United States, the concern has shifted to overnutrition and obesity and the risks faced by the developing fetal brain. Maternal obesity leads to serious inattention problems in offspring and a twofold increase in the incidence of impaired emotional regulation that was still evident five years after birth. Maternal obesity also causes abnormalities in areas of the brain responsible for feeding behavior and memory. All of these changes were most noticeable in male offspring. How does maternal obesity impair fetal brain development? Once again, the damage is due to the fact that fat cells release inflammatory proteins, called cytokines, into the body and brain of both mother and fetus. The more fat cells the mother has, the more cytokines get released into her blood. As discussed earlier, the presence of these cytokines increases the likelihood of becoming depressed. Exercise can modestly reduce the level of cytokines in the brain, and if overweight, moms might find some relief from their depression by exercising.

Why do I like to eat?

Two different neurotransmitter systems, endogenous opioid peptides called endorphins and cannabinoids, make eating pleasurable. Endorphins enhance the sensory pleasure derived from food, and the consumption of foods high in fat and sugar stimulates the release of endorphins. Endorphins enable us to experience the deliciousness of food and ensure that we do not stop eating too soon; endorphins do not influence our decision to eat. Drugs that selectively block the action of endorphins reduce the intake of foods that are quite sweet or have a high fat content. Interestingly, these drugs that block endorphins only reduce the pleasure of eating these foods; they do not reduce the feelings of hunger.

Endorphins drive us to overconsume palatable foods by blunting the impact of feeling full. As we all know, while standing next to the buffet table we will engage in mindless eating. We know that we should stop eating and move away from the buffet line and let someone else get at the food. Our bellies are full to the point that it hurts to breathe. Belts are loosened another notch. So why can't we stop eating? Neuroscientists have some interesting explanations. One of these is called "ingestion analgesia," and it involves endorphins. The function of ingestion analgesia is to keep you eating. Even though continued eating has become unpleasant because the stomach is painfully stretched to its full capacity, we still keep eating. Essentially, we block out the painful feedback from these feelings by releasing endogenous opiates into our brain and body. Not surprisingly, our reaction to pain is reduced significantly when eating tasty foods, such as cheesecake or rich chocolate. This explains why we can indulge in a decadent dessert even after we have become fully satiated by a large meal. We have basically become insensitive to the pain of continued eating. Also, if the animal eating next to you tries to take away your portion of the food, having your body flooded with endorphins will lessen the pain of any injuries that you sustain.

Brains evolved when food was scarce; thus, we are compelled by our genetic legacy to eat whatever and whenever possible until everything that can be consumed, is consumed to completion. All animals have a tendency to eat a great deal of food when palatable food is readily available. Not only that, but we also subconsciously prevent others from taking our food source. Just watch people's body posture at a buffet table. We defend our access to tasty food when it is within easy reach and is at risk of being consumed by other humans. Studies have shown that humans will eat more when more food is available even when the food is stale or otherwise unappealing (which is good news for bad cooks!). Furthermore, even if you point out to someone that the food is stale or that he has eaten more than his fair share, he will continue to eat. Our biological drive to consume tasty foods to completion outweighs any opposing cognitive or motivational factors.

Even after you have gained a lot of weight, your brain wants you to gain more. Research indicates that obese humans have elevated levels of endogenous endocannabinoids—marijuana-like chemicals—in the blood and brain. Remember "the munchies?" When we become overweight, our bodies induce a constant state of the munchies by bathing our brain in endo-cannabinoids. The endogenous marijuana neurotransmitters, the endocannabinoids, also contribute to the pleasure of eating. Scientists have discovered that marijuana increases the pleasurable response to eating sugar but has no effect on how much we dislike the taste of other types of foods. For example, if you hate eating peas or broccoli, smoking marijuana will not induce you to like eating them.

The ability of sugar to induce a rewarding feeling is caused by the release of dopamine in the brain's reward center. This brain region informs you that your brain likes this food and wants you to consume it more often. In the presence of marijuana, significantly more dopamine is released in response to the same amount of sugar-enriched food. So what does all of this mean? Your brain's endogenous marijuana system ordinarily modulates how good a particular food tastes; smoking marijuana simply enhances this natural mechanism in the brain. Your brain's main purpose is to help you survive and pass on your be-fearful-first genes. Eating is a critical and necessary behavior that the brain organizes and controls to allow daily survival. Therefore, the brain rewards itself for successfully consuming enough calories to survive by releasing these two powerful neurotransmitters—endorphins and endocannabinoids. Because of the manner in which evolution has shaped the response of our brain to food, overeating of calorie-dense foods has become a major health problem in the modern world. Our brains were shaped by evolution to be very efficient at instructing us to eat, but quite inefficient at stopping us from eating.

Why do I crave fat and sugar?

Fat and sugar are craved like heroin or methamphetamine. Why is this so? The answer is that these foods actually change how the brain functions. Day after day, year after year, the constant bathing of the brain in fats and sugar slowly changes how the neurons within our brain's feeding center behave. Along with these changes, gradual modifications in brain circuitry also occur; ultimately, your brain rewires itself to eat more fat and sugar every day in order to feed the ever-more-powerful new programing that is evolving inside your brain. Scientists once assumed that obese people were simply addicted to food in the same manner that someone becomes addicted to heroin— that is, food produces happy, pleasant feelings, and, therefore, eating lots of food would produce extremely pleasant feelings. Not so. A few years ago scientists discovered just the opposite was true: The brain's reward center decreased its response to eating tasty foods. In obese humans, dopamine function becomes significantly impaired in response to many years of poor diet. Consequently, obese people consume ever greater quantities of fat and sugar in order to mitigate the diminished rewards that were once experienced by consuming only one scoop of ice cream or a small donut.

Are we born destined to become obese?

For some people, apparently, the answer is yes. Environment, determined by both geographical and societal forces, plays an important role. The genes we inherit from our parents also play a role. Many studies have shown that children who have two obese or overweight parents are four times more likely to become obese themselves. To be considered low risk, the parents of the adolescents needed to be lean, with a body mass index less than 25. When the children in the high-risk group were shown pictures of tasty-looking, high-calorie foods, the dopamine-dependent pleasure centers in their brains became highly activated, especially as compared with the response of the same brain regions in the low-risk children. Some children who are destined to become obese apparently inherit a dopamine system that becomes much more excited at the sight of a chocolate milkshake than does the dopamine system in the brain of a child who is not destined to become obese as an adult. Then, in adulthood, the brain switches the rules and begins to require more fat, salt, and sugar in order to achieve a similar level of dopamine-mediated reward. Once again, take note of the fact that your brain has only one goal: keep you alive long enough to pass on your genes to your offspring. Once that has happened the forces of evolution no longer care about your survival. Thus, your brain will regularly induce you to consume foods that bring it pleasure regardless of the long-term health consequences.

How do my gut bugs keep my brain healthy?

Your brain lives in a symbiotic relationship with the bugs in your gut. Whatever you eat, they eat. In return, they help your brain function optimally in a variety of ways. During the past few years, it has become increasingly apparent that in the absence of bacteria humans never would have evolved to our current level of cognitive performance. Our brains are profoundly dependent upon a wide range of chemicals produced by these gut bugs. For example, without these gut microbes our brains do not correctly develop the serotonin neurons that play a key role in the control of emotion.

If you were to count all of the cells on and inside of you that are not actually *you*, they would number in the trillions. These bugs were not simply along for the ride as we became the dominant species on this planet; they made the journey possible. As soon as individual cells evolved into fully multicellular organisms during the Cambrian period about 500 million years ago, they quickly discovered the fantastic survival benefits of fully integrating themselves; once there, they never left. The total weight of the many trillions of bugs that reside in your gut is over two pounds and they are multiplying constantly thanks to all of the nutrients you are providing them; they are also in a constant battle for survival. The viruses in your gut kill so many bacteria every minute that their carcasses account for about 60% of the dry mass of your feces (now you know what is in there!).

Gut bacteria produce many different chemicals that can influence brain function. They convert the complex carbohydrates in our diet to the fatty acids butyrate, acetate, and propionate. Butyrate can easily leave the gut and enter the brain, where it can influence the levels of brain-derived neurotrophic factor (BDNF). BDNF plays a critical role in the birth and survival of neurons and the ability of the brain to learn and remember. Reduced levels of BDNF are correlated with impaired cognitive function and depression. Accumulating evidence suggests that gut bugs play key roles in both the developing and mature nervous system and may contribute to emotional and behavioral disorders as well as numerous neurodegenerative diseases.

Recent animal studies have shown that eating a high-fat diet can negatively alter the diversity of your gut microbiome, leading to reduced plasticity in the brain and increased vulnerability to anxiety. Eating a diet high in sugar also altered microbial diversity and significantly impaired learning and memory abilities.

Obviously, you need to take good care of these bugs so that they will take good care of your brain. Consuming prebiotics and probiotics can help us to maintain a healthy diversity within the bug environment. For example, elderly and frail humans who have major cognitive impairments also have the lowest level of bug diversity in their guts.

Can a good diet make you smarter?

Given that a poor diet can impair cognitive function, can a good diet make you smarter? Recently, a group of scientists investigated whether eating fruits and vegetables for 13 years would actually protect against a decline in cognitive abilities that humans commonly experience with normal aging. Their answer? Yes, it does; this is how they proved it. The study involved about 2,500 subjects who finished the study and adequately completed all the dietary and cognitive evaluations. The subjects were between 45 and 60 years of age at the beginning of the 13-year study, and they were required to maintain careful and detailed records of their daily diets. The subjects were evaluated at the beginning and end of the study for a variety of cognitive abilities, including verbal memory and higher executive functions such as decision-making and mental flexibility, along with numerous other tests. There is good news and bad news in the results. First, their diets were composed of a variety of fruits and vegetables, but specifically excluded potatoes, legumes, and dried fruits (each of these foods would have introduced specific complications that might have interfered with the outcome). The adults were divided into four groups according to the following diets: folate-rich diets containing fruits and vegetables, beta-carotene-rich diets containing fruits and vegetables, vitamin C-rich diets of fruits and vegetables, and vitamin E-rich diets containing both fruits and vegetables. The individual consumption of specific nutrients—folate, beta-carotene, and vitamins C and E—also was monitored. The subjects were allowed to choose how much of each diet they wished to consume each day; therefore, daily intakes of each nutrient varied. This was allowed in order to more closely reproduce how most of us actually select our daily intakes. At the end of the study, this is what they found. Eating fruits and vegetables has significant beneficial effects on very limited aspects of brain function. When the specific diets were examined more closely, diets that consisted of only fruits or diets with fruits and vegetables rich in vitamins C and E selectively benefited only verbal memory scores. This test involved being told to remember 48 different words and then recalling them after a delay with distractions.

The surprising finding was that eating fruits and vegetables had no significant benefit for other types of tasks that required alternative types of memory, such as learning motor tasks or recognizing familiar objects.

Clearly, each component of your diet may influence how well your brain works in unique ways. Natural antioxidants found in fruits and vegetables, like polyphenols, provide protective effects for the brain through a variety of biological actions. Polyphenols are everywhere in nature; more than 50 different plant species and over 8,000 such compounds have been identified in plant extracts. Obviously, investigating the multiple health benefits of these natural chemicals poses an enormous challenge. The most thoroughly investigated polyphenols are probably quercetin, which is found in apples, tea, and onions, and resveratrol, which is found in the skin of grapes. Grapes use resveratrol to defend against fungus. Tea contains a number of beneficial chemicals. In neurodegenerative diseases, administration of tea extracts reduced the production of mutant proteins and may prevent neuron cell death in Alzheimer's disease. Although tea is not a cure for Alzheimer's disease, its use is certainly justified given its safety and potential for long-term benefits.

What about an apple a day?

Are fruits good for you? After all, most fruits are full of sugar. Many popular diets recommend avoiding carbohydrates, especially sugar, in any form. There are some good arguments that could be made about avoiding sugar, but if this approach takes fruits out of your diet, you may be missing important nutrients that might make you healthier in the long term. One of these important nutrients is ursolic acid. Ursolic acid is found in apples (mostly in the skin), cranberries, and prunes, as well as in elderflower, basil, bilberries, peppermint, rosemary, thyme, and oregano. Eating fruits and spices that contain ursolic acid may enhance brain function and reverse some of the negative effects of obesity on the brain as you get older. Studies have shown that ursolic acid can improve cognitive functioning by increasing your brain and body's sensitivity to insulin. The biological mechanisms now have been investigated fairly well, and it appears that ursolic acid is able to correct some of the errors of metabolism induced by long-term obesity. The real challenge is to discover how many apples, prunes, and cranberries you need to eat in order to achieve these benefits; studies on humans have never been performed.

Will you lose weight by eating these fruits?

Maybe; it depends on what else you are eating. Will you lose weight avoiding fruits and berries while only eating meat? Yes. Over the long term, however, it is unwise to do so. The benefits of an all-meat diet are more immediate than the benefits of eating apples, cranberries, and prunes, because their effects on your health take longer to notice. Essentially, most of the restriction diets that are often promoted in popular magazines have not been around long enough for medical science to determine the long-term risks. Dietary restriction, which is reduced caloric intake without essential nutrient deficiency (i.e., a state of undernutrition without malnutrition), is the only valid, scientifically proven dietary intervention that has been shown to slow the aging process and improve health. There are two reasons why we hear so little about this approach: First, no one stands to profit from all of us eating less food and more apples, cranberries, and prunes. Second, the effects of dietary restriction on longevity have never been demonstrated in humans because rigorous and well-controlled clinical investigations have never been attempted. The effects of dietary restriction on health and longevity have been compellingly demonstrated across numerous species from single-cell organisms to rats to primates. If you are not willing to restrict your calorie intake, some alternatives are discussed in the following paragraphs.

Are spices good for my brain?

Cinnamon is a spice obtained from the bark of the *Cinnamomum verum* tree. Since antiquity, it has had many uses. Moses included it as an ingredient of the holy anointing oil. The Chinese knew it as Gui Zhi and recommended it for its antibacterial and antipyretic properties. Medieval physicians included cinnamon in their preparations to treat arthritis and infections. (The widespread use of willow tree bark [and the aspirin-like chemical that was derived from it] for these ailments was still a thousand years in the future.) A recent study found that eating cinnamon might prevent a variety of age-related neurological disorders. How does this happen? The sodium benzoate produced in the body after eating cinnamon induces significant increases in the levels of a variety of chemicals in the brain called neurotrophic factors. These factors stimulate the birth of new neurons in the brain and encourage the survival of existing neurons. These two processes are critical for the maintenance of a healthy brain. During the past decade, many scientific studies have discovered that these neurotrophic factors can prevent, or greatly slow, the progression of a variety of degenerative diseases of the brain, including Alzheimer's and Parkinson's disease. Cinnamon also can reduce blood sugar levels slightly in people with type II diabetes and reduce cholesterol levels by up to 25%. Thus, cinnamon is good for your brain and body.

Curcumin, derived from the spice turmeric, the powdered rhizome of the medicinal plant *Curcuma longa*, has been used for many centuries throughout Asia and India as a food additive and a traditional herbal remedy. Studies have shown that curcumin has potent antioxidative and anti-inflammatory proclivities that may be beneficial for patients with Alzheimer's or Parkinson's disease. Treatments with natural antioxidants and anti-inflammatories through diet or dietary supplements are becoming attractive alternatives. Epidemiological and research findings strongly indicate that the solution to healthy aging is exactly what you have heard from your mom: Eat healthy and in moderation; exercise in moderation.

> Let food be thy medicine and medicine be thy food.
> —Hippocrates (460–370 B.C.E.)

During the past 2,500 years since the time of Hippocrates, science has made significant progress in understanding how food exerts its beneficial effects on health. We now have solid proof that the foods and beverages that are consumed by humans, in particular those derived from tea leaves, coffee and cocoa beans, celery, grapes, mangos, berries, hops, and other grains, have clearly defined beneficial effects on brain function. Although these foods and drinks have quite different chemical compositions, they all contain compounds called flavonoids. Flavonoids are not in themselves nutritious, but they are believed to be responsible for the beneficial effects of many foods on the brain.

How do flavonoids benefit us?

In order to answer this question, scientists have investigated what flavonoids can do when their concentration in the brain is extremely low, that is, at levels that are likely achieved by a diet rich in these fruits. The flavonoids directly induce neurons in the brain to become more "plastic"—that is, more capable of forming new memories. The flavonoids achieve this by directly interacting with specific proteins and enzymes that are critical for learning and memory. They also induce the birth of new neurons, a process that is critical for recovery from injury, exposure to toxins, and the consequences of advanced age, such as increased levels of brain inflammation. Finally, some recent studies have shown that flavonoids actually enhance blood flow to active brain regions and thereby allow enhanced neuronal function.

So how much is enough? Let us consider two of our common favorites: wine and chocolate. If you consumed about

200 milliliters (6.7 ounces) of Cabernet Sauvignon or about 50 grams (1.7 ounces) of dark chocolate (71% cocoa powder), you would take in nearly identical quantities of flavonoids, which, fortunately, is now the daily wine intake recommended to produce the most health benefits in a typical adult. When young adult females were given flavonoid-rich chocolate drinks, blood flow to their brains significantly increased within just two hours, and their performance on a complex mental task greatly improved. No one is certain whether all flavonoids are capable of producing these benefits. Recent investigations have suggested that it does not matter which type of food provides the flavonoids, only that you should eat them as often as possible. In addition to those edibles mentioned above, studies to date also have identified benefits from black currants, pears, blueberries, strawberries, and grapefruit. You might have noticed that all of these healthy choices are darkly colored—their color is what makes them so valuable to your body. One final caveat: No studies have yet proven a true cause-and-effect connection between the lifelong consumption of flavonoid-rich diets and a reversal of age-related deterioration in learning or general mental function. Still, just in case, it might be worth modifying your diet accordingly (e.g., eating more chocolate).

Eat chocolate!

In 1648, according to the diary of English Jesuit Thomas Gage, the women of Chiapas Real arranged for the murder of a certain bishop who forbade them to drink chocolate during mass. In an ironic twist, the bishop was ultimately found murdered after someone had added poison to his daily cup of chocolate. Was this an act of blind rage by the women of Chiapas Real or justifiable homicide? For a small percentage of the population, eating chocolate can produce rage, paranoia, and anger that occur without warning. Fortunately, for most of us, this is not the typical reaction to eating chocolate. In order to understand why chocolate is so enjoyable for some while it induces uncontrollable rage in others, we need to consider the contents of most dark chocolates. Chocolate contains an array of compounds that contribute to the pleasurable sensation of eating it. Many of these compounds are quite psychoactive if they are able to get into our brains. Are they the reason we love chocolate so much? Are they the reason some people fly into fits of anger? The answer to both questions is, of course, yes. However, as is true for so many of the things we eat that affect our brain, it is not all that simple.

Chocolate usually contains fats that may induce the release of endogenous molecules that act similarly to heroin and produce a feeling of euphoria. German researchers reported that drugs that are able to block the actions of this opiate-like chemical produced by eating chocolate prevented the pleasure associated with eating chocolate. Chocolate also contains a small amount of the marijuana-like neurotransmitter called anandamide. Although this molecule can easily enter the brain, the levels in chocolate are probably too low to produce an effect on our mood by themselves. Chocolate contains some estrogen-like compounds, a fact that may explain a recent series of reports showing that men who eat chocolate live longer than men who do not eat chocolate. (The effect was not seen for women, who have an ample supply of their own estrogen until menopause.)

Let us focus on those women of Chiapas Real again. In contrast to its effects on men, women more often claim that chocolate can lift their spirits. In a study of college students and their parents, 14% of sons and fathers and 33% of daughters and mothers met the standard of being substantially addicted to chocolate. Women seem to have very strong cravings for chocolate just prior to and during their menstrual cycle. Women eat more chocolate in the days before the start of their period when progesterone levels are low. This is when premenstrual symptoms tend to appear as well. Chocolate may provide an antidepressant effect during this period. In one study, researchers found that women in their fifties often develop a sudden strong craving for chocolate. It turns out that most of the women had just entered menopause and were on a standard form of estrogen replacement therapy consisting of 20 days of estrogen and 10 days of progesterone. The chocolate cravings developed during the days on progesterone.

Chocolate contains magnesium salts, the absence of which in elderly females may be responsible for the common post-menopausal condition known as "chocoholism." About 100 milligrams of magnesium salt is sufficient to take away any trace of chocoholism in these women, but who would want to do that? Finally, a standard bar of chocolate contains as many antioxidants as a glass of red wine. Clearly, there are many good reasons for men and women to eat chocolate to obtain its indescribably soothing, mellow, and yet euphoric effect.

Okay, what about the anger? How might that happen? Chocolate contains phenethylamine (PEA), a molecule that resembles amphetamine and some of the other psychoactive stimulants. When chocolate is eaten, PEA is rapidly metabo-lized by the enzyme monoamine oxidase (MAO). Fifty percent of the PEA you consume in a chocolate bar is metabolized within only 10 minutes. Therefore, very little PEA usually reaches the brain, thus contributing little or no appreciable psychoactive effect. However, the amount of PEA in the brain might reach noticeable levels if MAO levels are low. Thus, it might not be a coincidence that MAO levels are at their lowest level in premenstrual women when they most crave the sooth-ing effects of chocolate.

Chocolate also contains small amounts of the amino acid tyramine. Tyramine can powerfully induce the release of adrenaline, increase blood pressure and heart rate, and pro-duce nausea and headaches. Usually, the nasty effects of tyra-mine are prevented because MAO metabolizes it, too. You can see the problem: The tyramine and PEA in chocolate may slow each other's metabolism. The consequence is that if both of these chemicals hang around too long in the body, high blood pressure, a fast-beating heart, heightened arousal, racing thoughts, anger, anxiety, and rage would ensue. One rather controversial study claimed that inhibitors of MAO were able to increase PEA levels in the brain 1,000-fold! That is a lot, and the consequences of this actually happening could be lethal. Nevertheless, the potential exists for some vulnerable people to experience significant shifts in mood after eating chocolate with high cocoa powder levels.

The main point to take away from this discussion about chocolate is that plants, such as the pods from the cocoa tree, contain a complex variety of chemicals that, when considered individually, are not likely to impact our brain function. When considered in aggregate, however, they may exert compound effects throughout the body; some of those effects may be desirable, while others may not. Chocolate is yet another excellent example of how difficult it is to differentiate food from drugs.

Brain toxins in the diet

Foods are full of toxins and nutrients; however, this is only from our human perspective, not the plants' perspective. The overwhelming percentage of toxic substances we consume exist naturally in the plants we consume. Indeed, most of the chemicals humans consider nutritious occur in the same parts of the plants that we consider toxic. Toxins and nutrients are byproducts of the plant living its plant life. We evolved the ability to defend ourselves from these toxins as long as the levels were not too high. Plants absorb lots of different molecules from the land in which they are growing. Some of these can interact with our brain. One famous example is aluminum. Aluminum is everywhere around us all of the time. It is the most abundant metal in the Earth's crust. Yet, somehow, we have become fearful of it when it is used as cookware, as cans for beer or sodas, or as deodorants. As far as anyone can currently determine, no plants or animals need it for any biological purpose. The reason is that aluminum is highly reactive and easily combines with other metals and oxygen to form hundreds of different minerals. Aluminum, in scientific terms, is not bioavailable to humans—usually. It all depends upon what chemical form the aluminum takes on. Usually, because aluminum is so tightly bound within minerals, animals have no chance to absorb it into their tissues.

This all changed a century ago due to the burning of certain types of coal for energy. In addition, anyone over a certain age will remember the fears associated with acid rain. Although the consequences of having elevated levels of sulfur dioxide and nitrogen oxides in the air have been known since the beginning of the Industrial Revolution, public awareness peaked in the 1970s due to the appearance of "dead lakes," the destruction of entire forests, and the pitting of marble statues in the United Kingdom and Europe. A century of acidic rains settled into the soil and changed the chemistry of minerals containing aluminum.

Plants do not use aluminum, but they are capable of absorbing it from the soil. Grains that are harvested today to make breads and cereals often contain a few parts per million of aluminum. However, the aluminum in grains unfortunately exists within a bioavailable form, that is, a chemical form that we humans can absorb into our bodies and deposit in tissues. Animals who eat these plants concentrate the aluminum in their tissues, too. Thus, meats obtained from cows may contain up to 1,000 parts per million of aluminum. This is where things get a little dicey. Are we at risk from the aluminum in our diet? As was true for drugs, it depends entirely upon how much one consumes.

Some people are vulnerable to the presence of aluminum in the body. For example, a few years ago people undergoing dialysis began using water containing high levels of aluminum. Over time the levels of aluminum in their brains and bodies began to increase and produced changes in their behavior that resembled dementia. The aluminum deposited within some brain cells caused those cells to die. Fortunately, dialysis centers are aware of this risk and have taken steps to prevent the problem from occurring again. Aluminum does not cause Alzheimer's disease, although it has been found in the brains of patients who have died with Alzheimer's disease. Although this seems suspicious, aluminum salts will deposit in any soft tissue that has cell loss due to injury or degeneration. For example, aluminum salts also will accumulate in the hearts of people with coronary disease.

What about deodorants? The aluminum salts used in these products do one thing—they irritate our sweat glands to the point that they swell and close the pores that allow perspiration to reach the surface of our skin. Essentially, aluminum prevents its own absorption by doing so. The real risk from deodorants comes from using sprays that produce a cloud of aluminum salts that can be inadvertently inhaled. Thus, keep using your aluminum cookware—it poses no risk to health.

4

WHY DO I SLEEP AND DREAM?

Our bodies are full of rhythms that are controlled by the brain. Why? A clue to the answer lies in the fact that these rhythms closely follow the rhythms of our planet. Our many biological rhythms exist and are so vital to our survival because we evolved on a spinning planet. For the past 3.5 billion years the most constant and reliable signal to all evolving plants and animals was the regular appearance of the sun on the eastern horizon. During the course of evolution, brains—no matter how simple or complex—always experienced a highly reliable pattern of light and darkness. We call this pattern of day–night cycling the circadian rhythm. Today, after 3.5 billion years of sunrises, our bodies have established a reliable routine: When the sun rises, your eyes inform your brain that it is time to prepare your body for the day's activities related to survival and procreation. Then, after the sunlight has receded in the West, and after a long day of being awake, your brain prepares you to fall asleep.

What is sleep?

What is it about being awake that leads to a deterioration of normal brain function? This chapter will explore why mental fatigue, poor attentional and decision-making capabilities, impaired learning, and a heightened risk of migraine headaches and epileptic seizures occur as a consequence of sleep deprivation. Sleep is critical to our overall health. Studies have shown that complete insomnia ultimately leads to death in humans, rats, and flies alike.

Each 24-hour day your brain has a choice of three highly distinct phases of consciousness that it can inhabit: (1) wakefulness and (2) dream sleep are periods of time when your brain is quite active; (3) nondream sleep (also called slow wave sleep) is a period when your brain's activity level is very low.

It is currently thought that the first appearance of the pattern of sleep that we humans are familiar with, that is, a cyclic balance between dreaming and nondreaming sleep, may have appeared in mammals about 150 million years ago. Prior to that ancient time, the brain was either in an active or inactive state. If this is true, the implication is that dinosaurs never dreamed. Dreaming required that the evolving brain find a way to achieve a good balance among time spent awake, time spent in nondreaming sleep, and time spent in dream sleep. For some animals, maintaining a good balance among wakefulness, dream sleep, and nondream sleep can be quite challenging. For example, seagoing mammals that must constantly be swimming or flying place one hemisphere into deep sleep while the other hemisphere remains wide awake. Dolphins, for example, must keep swimming and coming to the surface to breathe and thus can never place both sides of their brains into deep sleep.

Why is sleep so important?

What is so critical about sleep that your brain can justify shutting down consciousness? When brains were evolving, sleeping was a very dangerous thing to do. Being unconscious of your surroundings places you in danger of being discovered unaware and vulnerable to attack. Clearly, something about turning off your conscious awareness of the environment is critical to your survival, otherwise such risky behavior as falling asleep would not be permitted.

One recent theory states that with prolonged wakefulness the waste products produced by normal brain function lead to functional deterioration. According to this theory, sleep is required for the removal of these waste products of brain metabolism from the interstitial space among brain cells where they accumulate. In addition to providing an opportunity for the brain to flush itself of debris, sleep also is associated with many different changes in wiring between neurons so that the communication between these neurons is improved or enhanced. This change in the communication between neurons underlies the formation of new memories. Essentially, while you are sleeping your brain is reactivating patterns of neural activity that it experienced during the previous day while you were awake. The brain appears to rewind the videotape of the day's events, replay them, and delete memories that are weak, and possibly unimportant to you, and enhance memories that are strong and likely quite important to your survival. Essentially, sleeping and dreaming are ways that the brain has evolved to get rid of the chemical and mental debris that it collects during the day.

You do not "fall" asleep. Sleep is not a passive process. Transitioning from being awake to being asleep is an active process that involves the activation of many different neurotransmitter systems. The most critical neurotransmitter for the transition into sleep is gamma-aminobutyric acid (GABA), the brain's principal inhibitory neurotransmitter molecule. The amino acid neurotransmitter GABA always turns other neurons off. Drugs that enhance the action of GABA provide therapeutic benefits for a wide range of disorders, particularly for the treatment of anxiety and insomnia. For example, alcohol and many antianxiety drugs reduce brain activity by enhancing the actions of GABA at its protein receptors.

What other daily rhythms do I experience?

Each cell and organ system in your body follows a daily rhythm that is initiated by the rising sun every morning. One of the most important rhythms is body temperature. You are a warm-blooded animal, and your brain carefully defends you from cooling down to match the temperature of your environment. Your body temperature fluctuates rhythmically throughout the day. Every day you wake up cold and quickly become warmer, reaching a plateau in the morning that you maintain throughout the day. At the end of the day, you start cooling down again as you fall asleep. This is why it is best to keep the bedroom as cool as possible; at the end of the day, it assists with your transition into normal sleep.

While you are sleeping, the nighttime is full of important rhythms, too. The onset of sleep is associated with many significant fluctuations in various hormones. When you fall asleep, your body releases growth hormone and other wound-healing chemicals; after all, the brain has learned that for the next few hours it has a chance to heal. Your body's production of cholesterol and triglycerides increases just after you fall asleep, which explains why statin cholesterol-lowering drugs are most effective if taken at bedtime. Blood levels of the stress hormone cortisol are surprisingly high when you wake up in the morning and then gradually fall throughout the day. The rhythmic clocks in your brain also influence how you feel, how you think, and how drugs affect your brain. You perform mental tasks much better in the late morning and you are more vulnerable to the disruptive and toxic effects of alcohol and anesthetic drugs in the late evening.

What happens when I disrupt these rhythms?

Traveling across time zones is disruptive to brain and body performance, especially when flying in the same direction as the earth is spinning, namely, east, because then you experience the sun rising sooner than your brain expects. For example, visiting teams who fly east win an average of 37% of their games; visiting teams who fly west win an average of 44% of their games; obviously, it is best to be the home team. Scientists have learned that we need to maintain normal rhythms of waking each day and falling asleep each night at the same time; if we do not, the consequences can be far more negative than just losing to the home team.

To study the impact on your brain and body of losing these normal rhythms, watch what happens when you isolate yourself from the influence of morning sunlight, for example, by living in an underground cave. At first, your major rhythms are maintained; you go to sleep at roughly the same time, become colder as the night progresses, and awaken at the usual times. Then, as you spend more time underground, without noticing it you start going to bed about one hour later every night. Why? Because, like most people, you do not have a true 24-hour clock in your brain. In fact, most of us are born with a 25-hour clock in our brains. (Some people may even have 36-hour clocks!) This is why everyone finds it so easy to stay awake one extra hour to watch a favorite television program and why it is so very hard to go to sleep earlier than usual.

In addition to falling asleep one hour later each day, because you do not have access to the morning sunshine, your body begins to make other adjustments. One major change involves the rhythmic fluctuation in your body temperature, particularly the point at which you are the coldest each day. Before moving into the cave you would have experienced your daily minimum temperature early in the morning, just before waking. Now that you are living in a cave you start feeling colder at mid-day, and then at the beginning of the day. These alterations in essential biorhythms indicate that your body is undergoing some very significant disruptions. Unfortunately, these disruptions in your biorhythms have lingering effects after you emerge from your underground home. One of the most common negative consequences is depression. Current thinking is that people who are vulnerable to depression might have their illness triggered by disruption of their healthy sleep-wake pattern. Alternatively, recent studies of the genes involved in controlling the brain's circadian clock mechanism suggest that inherited abnormalities in these genes may underlie mood disorders such as bipolar illness. This is why it is so important for people with major depressive disorder, or bipolar disorder, to maintain an inflexible sleep–wake pattern; that is, these patients must go to bed at the same time every night. Disruption of the normal sleep–wake pattern also might underlie why traveling across multiple time zones can induce a depressive episode in vulnerable people.

How are eating and sleeping related?

What you eat before bedtime also might improve your chances of getting a good night's sleep. A recent study suggests that eating something sweet might help induce drowsiness. Elevated blood sugar levels have been shown to increase the activity of neurons that promote sleep. These neurons live in a region of the brain that lacks a blood–brain barrier; thus, when they sense the presence of sugar in the blood, they make you feel drowsy. This might explain why we feel like taking a nap after eating a large meal. This is just one more bit of evidence demonstrating your brain's significant requirement for sugar in order to maintain normal function.

During the day, fat is deposited, muscles increase their metabolism of fats and sugars, your liver is busy producing glycogen (a form of sugar storage) and bile (in order to absorb fats), while your pancreas is busy releasing insulin in response to eating. At night, these processes reverse; for example, fat catabolism is increased. This explains why getting adequate amounts of sleep, that is, seven to eight hours, is so important for the maintenance of a normal body weight. While you were sleeping and dreaming, your brain and body used up a lot of energy; thus, as soon as you awaken, your body needs to replenish itself.

How does your brain control its rhythms?

The sun has arisen and you are awake; what are you going to do now? Eat.

These two events have occurred together for as long as your brain has been evolving on this planet. Thus, it is not at all surprising that your brain uses the same neurotransmitter molecule, called orexin, to both arouse you from sleep and then induce you to eat. There are only about 70,000 orexin neurons in the entire human brain; however, these relatively few neurons control many other better known neurotransmitter systems, such as serotonin, histamine, and acetylcholine, whose roles are to control your level of arousal and attention throughout the day and night. You should view orexin as the one ring to control them all.

What would happen if your orexin neurons started dying? The answer is that you would develop narcolepsy, an incurable neurological disorder. Narcolepsy is an autoimmune disorder, which means that your body is attacking a part of itself; in this case, the orexin neurons in your brain are being attacked by your body's immune system. The symptoms of narcolepsy are excessive daytime sleepiness, sleep paralysis, hypnagogic hallucinations (that occur at the onset of sleep) and disturbed nocturnal sleep. People with narcolepsy experience intermittent, uncontrollable episodes of falling asleep during the daytime; these symptoms are sometimes evoked by strong emotions. People with narcolepsy begin dreaming almost immediately upon losing consciousness. This is unusual because typically most people do not start dreaming for almost 120 minutes after falling asleep. Recent studies suggest that the loss of orexin neurons, long-term stress, and sleep deprivation may be associated with obesity and age-related decline in cognitive abilities. This connection will become more obvious as the details of sleep are discussed below.

Sleep is composed of two distinct phases: non-rapid eye movement (non-REM) sleep and rapid eye movement (REM) sleep. Non-REM sleep has four unique stages, ranging from light to deep sleep. After initially falling asleep, you spend about 90 minutes sequentially passing through deeper and deeper stages of non-REM sleep; then you reverse this process and slip into lighter states of non-REM sleep before finally switching over to your first REM sleep episode. REM sleep is the most mentally active stage of sleep and is when dreaming most often occurs. During REM, or dream sleep, your eyes move back and forth beneath the eyelids; this is why the term rapid eye movement sleep was given to this stage of sleeping. Your brain experiences about six to eight non-REM-to-REM cycles of sleep during a typical night of sleep. Normal amounts of REM sleep facilitate creative problem solving. Thus, night-time dreaming is a necessary component of making us smarter during the day. It is not, however, the total quantity of sleep that you have each night that is so critical; neuroscientists now understand that the quality of sleep each night (i.e., the balance between non-REM and REM sleep phases) underlies optimal brain function during the daytime.

Why do I sometimes wake up paralyzed?

During REM sleep, your muscles are actively paralyzed so that you do not act out your dreams and risk discovery by predators while you are vulnerable. During dreaming, the muscles of your body are inactivated by a complex interaction of many different neurotransmitter systems descending into the spinal cord. Sometimes as you transition into, or out of, your dream state, this paralysis can turn on too late or not turn off in time. This is called "REM atonia" and is experienced by about 60% of the population at least once in their lives. Waking up paralyzed is likely due to the failure of GABA-containing neurons in the brain to turn off. In a typical event, people report that they are fully awake but unable to move. Naturally, this causes considerable fear. Many people also report that they feel as though they are not alone in their bedroom when this happens—which adds to their fear. Typically, the paralysis disappears quickly without any lingering problems. Frequent wakening paralysis may indicate underlying problems with brainstem function; however, too little information is available at the current time to understand fully the reasons behind wakening paralysis.

Why do some people act out their dreams?

Some sleepers, usually elderly men, make abnormal flailing movements during sleep while acting out their dreams. These movements are potentially dangerous to their bed partners, who often suffer injuries inflicted by their larger male mates. This syndrome is call REM Behavior Disorder and is likely due to the degeneration of a group of GABA-containing neurons in the brainstem. Middle-aged men and women who demonstrate REM Behavior Disorder are at increased risk (by about 50%) of developing Parkinson's disease about six years after this symptom appears. No particular lifestyle or nutritional behaviors, such as drinking caffeinated beverages, smoking, or consuming alcohol, altered the onset of symptoms. Recent evidence suggests that the symptoms of REM Behavior Disorder may occur quite early in life, possibly even during the second decade of life.

What happens when I am dreaming?

Everyone dreams about something. Even if you do not remember your dreams from last night, you did dream. It was once believed that you only dreamed during REM sleep. Now scientists realize that not all dreaming occurs during REM sleep; you also dream during non-REM sleep, although the nature of the dreams in each stage is quite different. What is your brain doing when you are dreaming? This is what we know from studying animals and humans when they are dreaming: During the day, your hippocampus was busy gathering sensory information associated with the events of your life. At night while you are in non-REM sleep, the hippocampus "shows movies" of the day's events to the frontal cortex. These "movies" are presented in very compressed packets of neural information. The "movies" are presented about 11 times faster than the events actually occurred in real life. Why? Just consider the challenge your brain faces every night—lots of things happened during the previous 18 hours and all of this information needs to be processed within the much shorter period of time that you spend in REM sleep. Thus, because you spend much more time awake gathering information and knowledge than you spend asleep processing this information, your brain needs to work very hard while you are dreaming. In spite of this speedy processing by our neurons, most dreams have a realistic sense of time (more on this later). Dreams that occur during non-REM sleep tend to be simpler and have less narrative quality as compared with the more complex, and usually more emotional, dreams that occur during REM sleep.

Dreams are about the people you know, places you have visited, places you have seen on television, or in movies, events that occurred in the books you have read, creations of your imagination, and the everyday normal events of your life. These thoughts and memories are the alphabet that writes the storyline of your dreams. For example, I am fairly certain that no one reading this page has ever dreamed of meeting me (unless, of course, you have met me). Furthermore, before the appearance of the Star Wars movies, no one ever dreamed of a Wookie or a Tatooine Canyon Krayt dragon. No one has ever dreamed of meeting an actual alien from another planet simply because no one on this planet has ever met one.

The content of men's dreams differs in important ways from that of women's dreams. When hundreds of test subjects were awakened while dreaming, the women frequently reported seeing bright colors while the men reported seeing very few bright colors; most objects for men appear in washed-out pastels or shades of gray. Women often report seeing brighter colors during menstruation. Women also reported knowing the identity of the sexual partners they dreamed about and noticing their partner's hands and face. In contrast, the men usually did not know the identity of their sexual partners and did not report seeing their partner's face. Pay close attention to your own dreams, when you remember them, and notice how your dreams compared with these reports.

Most dreams tend to emphasize strong emotions at the expense of reason. Neuroscientists believe that the presence of strong emotions while dreaming is related to the fact that a specific brain structure, the amygdala, is quite active during dreaming. What you fear while dreaming is likely also what you fear while awake. Your personality does not change when you are dreaming. Dream content reflects your basic waking conceptions; thus, if you are a devout Christian or Republican when you are awake, you will maintain similar views while dreaming.

Why do I sometimes dream that I am being buried alive?

Why do some dreams involve the terrifying feeling of being buried alive or the feeling that it is difficult to breath? These dreams of suffocation usually occur during non-REM sleep when your respiration and heart rate are slowed down significantly. If you are dreaming while experiencing these physiological conditions, your brain incorporates their sensory qualities into your dream narrative. Sometimes, just being wrapped up in your bed sheets provides a sufficient sensory stimulus to induce a dream of suffocation.

Are children's dreams different?

The dreams of children have not been studied as extensively as those of adults. In general, the dreams of children differ from those of adults and greatly depend upon the brain's level of maturation. Children usually are first able to talk about their dream experience around the age of two years. Only 20% of the time do children under eight years of age awaken from REM sleep recalling dreams (as compared with 90% of adults). Some studies suggest that the content of young children's dreams is often static and bland, not animated or emotional in the way that most adults experience dreams. Too often, however, these studies are confounded by the fact that the children were not fully aroused when questioned about their dreams. The dreams of a child do not regularly include the dreamer as an active character until after six or seven years of age, although some children report being an active participant as early as three years of age. This suggests that the appearance of dreaming, similar to many other higher cognitive abilities in young children, is due to a gradual, and highly variable, developmental process. The dreams of young adolescents, particularly in the case of recurrent dreams that tend to be more unpleasant, frequently involved confrontations with monsters or animals, followed by physical aggression, falling, and being chased.

What is a lucid dream?

For the majority of people, dreams are typically not volitional, that is, not under their direct control, nor do they feature self-reflection, insight, judgment, or abstract thought. Some people, however, are able to control the content of their dreams—we call these people lucid dreamers. How do the brains of lucid dreamers differ from the brains of nonlucid dreamers? When nonlucid dreamers are dreaming, the frontal lobes of their brains remain relatively inactive. In contrast, when lucid dreamers are dreaming, the dorsal, or top, part of the frontal lobes of their brains are quite active. Neuroscientists believe that this difference in the activation pattern of the frontal lobes underlies the ability of lucid dreamers to be self-aware during their dreams and to be able to adjust their dream narrative. Lucid dreaming is much more common among children and adolescents than among adults. This suggests that the loss of the ability to experience lucid dreaming is related to maturation of the brain.

The ability of some people to recall their dreams better than other people also may be related to which brain regions are active while they are dreaming. People who recall their dreams better have more blood flow within their medial frontal and posterior temporal lobes while dreaming than do people who do not recall their dreams as well. Thus, remembering dreams depends upon whether or not the appropriate brain region was paying attention (i.e., was active) while you were dreaming.

How long do dreams last?

While you are dreaming your brain does have the ability to estimate the passage of time without referring to time cues. The accuracy of this function fluctuates, however, from an overestimation during the initial hours of sleep to an underestimation during the final hours of sleep just before waking. Why? Time flows faster during non-REM sleep, which predominates during the initial phases of sleep. Thus, dreams during non-REM sleep are associated with longer estimates of sleep time subjectively experienced. In contrast, REM sleep predominates during the early morning hours. Thus, REM sleep is associated with longer estimates of sleep time subjectively experienced. Therefore, during the few hours just before your alarm sounds in the morning, your dreams seem to take a lot longer to unfold.

Does it matter what time I get up or go to bed?

Yes. Morning types get up early, perform mentally and physically best in the morning hours, and go to bed early. Evening types prefer to stay out late, get up at a later time, and perform best, both mentally and physically, in the late afternoon or evening. Evening-type individuals were significantly more likely to suffer from poor sleep quality, daytime dysfunction, and sleep-related anxiety as compared with morning-type individuals. Even more disconcerting is that late bedtime is associated with decreased hippocampal volume in young healthy subjects. Shrinkage of the hippocampus has been associated with impaired learning and memory abilities. So yes, it probably does matter what time you prefer to get up and go to bed.

Does it matter who I sleep with?

Yes, very much. Couples sleeping in pairs were investigated for sleep quality, that is, for the correct balance of non-REM and REM, as well as their own subjective view of how they slept. For women, sharing a bed with a man had a negative effect on sleep quality. However, having sex prior to sleeping mitigated the women's negative subjective report, without changing the objective results; that is, her balance of non-REM and REM was still abnormal. In contrast, the sleep efficiency of the men was not reduced by the presence of a female partner, regardless of whether they had sexual contact. In contrast to the women, the men's subjective assessments of sleep quality were lower when sleeping alone. Thus, men benefit by sleeping with women; women do not benefit from sleeping with men, unless sexual contact precedes sleep—and then their sleep still suffers for doing so.

Why do I need that morning cup of coffee?

The alarm rings, you awaken, and you are still drowsy: why? Being sleepy in the morning does not make any sense; after all, you have just been asleep for the past eight hours. Shouldn't you wake up refreshed, aroused, and attentive? No, and this is why. During the previous few hours before waking in the morning, you have spent most of your time in REM sleep, dreaming. Your brain was very active during dreaming and quickly consumed large quantities of the energy molecule ATP. The "A" in ATP stands for adenosine. The production and release of adenosine in your brain is linked to metabolic activity while you are sleeping. There is a direct correlation between increasing levels of adenosine in your brain and increasing levels of drowsiness. Why? Adenosine is a neurotransmitter that inhibits the activity of neurons responsible for making you aroused and attentive. You wake up drowsy because of the adenosine debris that collected within your brain while you were dreaming. The cure for your morning drowsiness: coffee. The caffeine in your coffee blocks the actions of adenosine and releases your neurons from their chemical shackles; your attentiveness improves and you are ready for anything—at least until the effect of caffeine wears off.

Caffeine is the single most widely consumed (legal) psychoactive ingredient in the United States. It is easy to understand why. It quickly enhances our physical and cognitive performance and usually improves our mood. Numerous studies have shown that caffeine improves performance on attentional tasks. Unfortunately, most of these studies were performed on young adults; neuroscientists know very little about the potential benefits of coffee in older adults. Some studies have found no benefits while others have concluded that older adults actually benefit more from caffeine than do younger adults.

Can I get too much sleep?

Do you remember when you were a child and had the measles, or some other infection, and you fell asleep on Monday and woke up on Thursday? When you are sick, elevated levels of invading bacteria or viruses induce your immune system to defend you. The debris from this battle includes a collection of molecules released from the carcasses of these invading bugs, including pieces of bacterial cell wall (called lipopolysaccharide) and bits of nucleic acids from inside the viruses. This is all part of your normal immune response. As a consequence, these inflammatory proteins floated into your brain and induced a normal and very prolonged sleep cycle. Your brain evolved to respond to illness by making you sleep in order to allow your injured or infected body to heal.

This same principle is true for other animals. Animal species that sleep more tend to be healthier. The daily number of hours an animal sleeps is correlated with the total white blood cell count and inversely correlated to the relative frequency of infections experienced by that animal species. Sometimes, however, the normal response of the brain to induce sleep goes too far and sleep can last too long.

Kleine-Levin syndrome is a good example of this extreme response by the body. Occasionally, following a minor flu-like illness with upper airway infection and an acute mild fever or even tonsillitis, some people will experience attacks of excessive sleeping, called hypersomnolence, which usually appear and end quite suddenly. Sometimes sleep lasts from several days to several weeks. The interval between attacks can last for several weeks to months, and sometimes even many years. Thus, getting too much sleep is not always an indication of good health.

What if I do not get enough sleep?

Although scientists have not discovered why we sleep, they have discovered that we need between six and eight hours every night. Not getting enough sleep makes us more likely to pick fights and focus on negative memories and feelings. The emotional volatility is possibly due to the impaired ability of the frontal lobes to maintain control over our emotional limbic system. We also become less able to follow conversations and more likely to lose focus during those conversations. Sleep deprivation impairs memory storage and also makes it more likely that we will "remember" events that did not actually occur. Extreme sleep deprivation also may lead to impaired decision-making and possibly to visual hallucinations. Not getting enough sleep on a consistent basis places you at risk of developing autoimmune disorders, cancer, metabolic syndrome, and depression. Why? Some recent studies have reported that sleep is important for purging the brain of abnormal, and possibly toxic, proteins that can accumulate and increase the probability of developing dementia in old age. Whatever you are doing right now, stop and go take a nap. When you get older, you will be glad that you did.

What happens to sleep with aging?

Normal sleep becomes quite disturbed with normal aging. First of all, it takes you a lot longer just to fall asleep; in addition, you are going to experience more nighttime awakenings. This leads to a greater tendency for daytime sleepiness and the need for short naps. Part of the problem is that older humans enter the deeper stages of sleep less often. This causes some health problems because it is during the first deep sleep periods that your body releases hormones necessary for healing and growth. The consistent loss of these deeper stages of sleep may contribute to slower wound healing and increased vulnerability to diseases that often fall into the category of age-related disorders. The aged brain also spends less time in REM sleep. The consequences of reduced REM sleep likely underlie age-associated problems with learning and memory.

Insomnia and other sleep disturbances are common in patients with neurodegenerative disorders, such as Alzheimer's disease and other dementing disorders. Disturbed sleep patterns, including chronic insomnia, represent a significant public health concern that is relatively common across all cultures.

Insomnia is possibly the first sign of aging in humans. Sleep problems that are similar to those seen in the elderly first appear just after puberty. Most people experience their deepest, restive sleep around the age of 10; then, sleep quality begins to decline. Sleep problems are associated with a poorer quality of life, as well as with mental and physical health problems. Does physical activity help us sleep better? An analysis of over 50 recent studies found that regular exercise, regardless of exercise intensity or aerobic classification, had a rather moderate benefit on overall sleep quality, and only a small positive benefit on total sleep time and sleep efficiency. So, if a little exercise only offers a modest benefit for sleep, what else might help? Low doses of melatonin also may be an effective option for the treatment of insomnia. Sadly, typical over-the-counter sleep aids are only modestly effective, quite addictive, and ultimately lose their effectiveness within a few days. In addition, all of the over-the-counter sleep medications are common antihistamines that do not reproduce normal sleep patterns. That is why your sleep is not restful when using these medications. It is usually best to avoid these medications, particularly as you get older.

5

HOW DOES THE BRAIN AGE?

This is an essential question to answer because the brain is a key regulator of your lifespan. Your body ages almost as fast as your brain; thus, anything that you do to slow the aging of your body will be reflected in a slower aging of your brain. Many of the brain's functions discussed in previous chapters, such as sleeping, learning and memory, eating behaviors, and emotional stability, may change significantly during normal and pathological aging. The nature and severity of these changes depend upon numerous factors, including diet and drug use, sleeping habits, and inherited vulnerabilities. During the past decade, it has become clear that the body's normal repair processes are programmed by your DNA to decline or are forced to decline due to your lifestyle choices.

How fast you age is almost entirely related to how your body generates the energy that your brain requires to function. Is there any way to slow this process down? Yes, there is; the best advice anyone could possibly provide was offered over 2,400 years ago by Hippocrates of Kos:

> If we could give every individual the right amount of nourishment and exercise, not too little and not too much, we would have found the safest way to health.

Alternatively: All things, including food and exercise, in moderation. Apparently, humans were overindulging even two millennia ago.

When does age-related senescence begin?

This is not an easy question to answer. Aging is hard to define; scientists usually define it operationally depending upon what feature of aging they are studying. People age at vastly different rates for quite different reasons: it all depends upon one's lifestyle. Did you smoke? Were you obese? These two factors are responsible for most of human aging and poor health. In addition, one's age according to the calendar does not necessarily correlate with a decline in all cognitive abilities. For example, as you age, your vocabulary will continue to improve and you might cultivate some subtle social skills. The earliest true sign of aging is a decline in sleep quality and the consequent impaired learning, memory, and attentional abilities during the day. As mentioned in Chapter 4, the release of adenosine in your brain is linked to metabolic activity while you are sleeping. Older brains have more difficulty rinsing out extracellular adenosine, thus leading to impaired learning and attention.

The second age-defining change is heralded by a reduction in your ability to experience strong emotions. Overall, with normal aging, the brain makes numerous compensatory changes in how information is processed in order to allow mental activity to remain somewhat normal. This chapter will begin by examining what happens during brain aging and then conclude with a discussion of what has been proven scientifically to slow brain aging. First, let us skip to the end of the story.

When am I going to die?

Many factors determine brain aging and, therefore, influence when you will die. Some of these factors have good explanations; others do not. The month in which you were born is related to how long you live. In the northern hemisphere, people born in May or December live longer than people born in February or August. You are far more likely to die in the months of January or February than in July or August. These statistics are biased due to the fact that most of the world's population lives in the northern hemisphere of the planet; thus, most of us experience cold winter nights in January and February.

Most people who die in their sleep usually die during the early morning hours when their body temperature is lowest, and it is easiest to become colder at night in the winter months. Do we really get so cold at night that we are at risk of dying? Yes. How is this possible, given that we have no problem staying warm during the day? The problem is that during REM sleep, the time that you are busy dreaming, your highly evolved mammalian brain does something truly bizarre: it reverts you back to the physiology of a lizard and you become a poikilotherm. Poikilotherms are animals, such as fish, amphibians, and reptiles, whose internal temperature varies in parallel to the ambient temperature of their environment. Therefore, because you spend so much time in REM sleep during the early morning hours, your body temperature is not defended by your brain and you slowly cool down. This is why in the morning you wake up feeling cold: you are cold.

Consequently, due to the fact that your body temperature is falling throughout the early morning hours, considerable stress is placed upon an aging cardiovascular system. Hence, elderly humans with failing cardiovascular function, an all too common condition today, tend to die during the early morning hours; you can chalk this up to your evolutionary link to reptiles. As you will read later, however, sometimes being just a little colder has significant advantages.

How can I live longer?

Some people live very long lives. What did they do differently from their contemporaries who died before them? Scientists have discovered some of their secrets by interviewing centenarians. Humans who have lived to be more than 100 years old consistently report three important lifestyle features: they lived most of their lives in the same place and near family members, they did not use tobacco products, and they used as few medicines as possible. In addition, being from a higher economic status, obtaining a higher education that leads to a nonroutine profession, being female, and having long-lived parents all correlate with a longer life.

Females of all species, whether flies, beetles, rats, spiders, or humans, live longer than males. Why? There may be a relation between telomere length and age. Telomeres are critical pieces of our DNA; every time a cell divides, the length of the telomere shortens. Women exhibit less telomere shortening over time than do men. Recent evidence suggests that this accounts for the longer lifespan of women. In addition, the presence of testosterone in men makes them age faster. To understand why men always die sooner than women, you also must understand the effects of eating.

How does eating age me?

Like most other animals on this planet, humans acquire energy for our biochemical machinery by breaking down the carbon bonds found in fats, sugars, and proteins, and then gobbling as much energy from the process as possible. Recent studies have discovered that humans, probably due to our massively active brains, metabolize food much faster than other animals. Much of the energy in our food is lost to heat that helps to maintain our body temperature. The process of extracting energy from the fats, proteins, and carbohydrates in our diets leaves our cells with leftover carbon atom debris. Think of this carbon debris as similar to the ash left over after a fire has consumed a piece of wood. This carbon atom waste must be discarded somehow. The solution for your cells, a solution that evolved at least 3.5 billion years ago, is to combine these leftover carbon atoms with oxygen. Your cells simply expelled this waste product as a gas called carbon dioxide: one carbon and two oxygen atoms bound together. Voilà, the problem was solved: carbon bond energy is consumed in the form of fats, carbohydrates, and proteins. Our bodies then extract energy and excrete the residue as carbon dioxide and water vapor. There was a big problem, however, with this ancient solution: oxygen is exceedingly toxic to our cells. Oxygen must be transported inside your body very carefully because oxygen causes oxidation; you might know this effect as rusting. Your body's solution for handling such a toxic molecule is to bind oxygen to a protein in the blood called hemoglobin.

Overall, the hemoglobin in your blood does a decent job of regulating the oxygen levels near the individuals cells of your body so that your cells have the oxygen they need for respiration, that is, for the removal of carbon waste, but do not have so much as to kill them outright. Sometimes, however, a stray oxygen molecule does get loose from hemoglobin and must be captured before it can do any real damage to your cells or their precious cargo, the DNA.

Cells also have evolved numerous antioxidant systems to defend you from the oxygen that you must breathe in; these antioxidant systems are so effective that they will allow you to live to be 100 years old, or older, if you are lucky. All species have had to deal with this challenge; the lifespan of virtually all species is highly correlated with how well they defend themselves from oxygen. Therefore, the best way to age slowly, and also be healthier while aging, is to expose your cells to as little oxygen as possible. One proven way to accomplish this is to need to eat as little food as possible. The most recent evidence suggests that you should focus your efforts on reducing the number of calories obtained as animal protein. The evidence to support this recommendation is overwhelming.

During the past few decades, studies have shown that the single most important factor that determines when you will die is how long you have been alive. That sounds obvious, but it begs the question: what do you do *every day* that increases your chances of dying? The answer for all humans, as well as for every other respiring animal on the planet, is simple: we eat and breathe. If you consume fewer calories, namely, carbon bonds in the form of fats, proteins, and carbohydrates, you require less oxygen. However, we must eat and breathe to survive; the problem is that doing so makes us vulnerable to the consequences of oxygen. Consequently, our bodies and our brains age more rapidly because we keep eating and eating and breathing and eating and breathing all day, every day of our lives.

Currently, scientists believe that restricting total calorie intake will not allow you to live longer; rather, reducing the total number of calories consumed each day by approximately one third will promote a healthier longevity and better quality of life.

How does caloric restriction work?

With normal aging, because you are eating and breathing, tissue-damaging molecules called oxygen free radicals are formed by the oxygen in your blood. Free radicals become more prevalent with age and may slowly overwhelm your natural antioxidant systems, destroying your neurons and just about every other cell in your body. According to another recent discovery, the overproduction of these oxygen free radicals may encourage cancer cells to metastasize and move around your body. Think about the unbelievable irony of this process: The biochemical processes that occur in every cell of your body are actively injuring those cells by the very process of trying to keep them alive. It turns out that each species' maximum lifespan may be determined by how many free radicals are produced in each of their cells. To paraphrase Pogo, we have met the enemy and he is in us. There are intimate, complex, and not well-understood relationships among the actions of mitochondria within our cells, our general health, and how fast we age.

This is worse for males. The combination of more muscle mass than women and the presence of testosterone makes men warmer than women. Testosterone alters how males metabolize food and increases the amount of heat their muscles produce during normal respiration. Testosterone turns the normal food-to-energy conversion process inefficient, that is, cells waste more energy as heat to make men feel warm. This is why it is so much easier for men to lose weight than for women; the male body, particularly the muscles, is capable of wasting a considerable number of consumed calories as body heat. Wasting calories to produce heat has negative consequences. Men need to consume more calories per day and thus produce more harmful oxygen free radicals than women do. Lacking both testosterone and significant muscle mass, women tend to produce less body heat from their food; consequently, it is much harder for women to lose weight than it is for men. In contrast, women waste less energy, need to consume fewer calories, and produce fewer oxygen free radicals—all of which benefit their overall longevity.

In summary, because you must consume food and breathe oxygen to survive, you slowly age and die. Oxygen is the troublemaker in our lives. Species that produce fewer free radicals from the oxygen they breathe, or that have evolved better ways to handle these toxic molecules, live longer than species that have not evolved these methods. There is a strong correlation between dying and how long you have been alive. What underlies this correlation? What are you doing every day of your life that ages you? Eating and breathing. Do you have any other recourse to slow your aging process? Yes, consume foods that remove the oxygen free radicals. For obvious reasons these foods are called antioxidants.

Antioxidant-rich foods include colorful fruit and vegetables, fish and olive oils, fruit juices, anti-inflammatory plants and drugs such as aspirin, some steroids, cinnamon and some other spices, nicotine, caffeine and chocolate, the fat-soluble vitamins, nuts, legumes, beer, and red wine. People who eat these foods do not report acute changes in their thoughts or moods (depending upon how much they consume!) but certainly benefit from consuming them regularly over their lifespan. In general, the benefit comes from the fact that all of these foods provide your brain and body with protection against the most deadly thing we expose ourselves to every day—oxygen. Thus, the lifestyles of people who live the longest are characterized by the consumption of foods rich in antioxidants as well as the consumption of far fewer calories overall. Both of these dietary approaches are healthy; when combined, they offer the best chance of a healthy life and slower aging.

Which brain region suffers the most with aging?

An unhealthy diet has many negative long-term consequences for brain function as we age. One of these is degeneration of the hippocampus in the temporal lobe, leading to the symptoms of dementia. There are two major consequences of the degeneration of the hippocampus: one is wandering and the other is memory loss. Sixty percent of elderly humans with dementia will wander; they forget their own name or address and become disoriented even in familiar places. Why? Recall from Chapter 1 that the hippocampus forms maps and informs you of where you are and where you live and where you parked your car. Aging alters hippocampal function; consequently, mental maps do not form correctly. Studies on mice, rats, nonhuman primates, and humans have allowed scientists to investigate what changes occur in the hippocampus during normal aging, as well as when and how they occur.

Young rats, similar to young humans, create a map of their immediate world and store it for future use. If they find themselves in this familiar environment again, they go to the correct map drawer in their brain and use it to find their way around. In contrast, when aged rats find themselves in a familiar environment, they either fail to retrieve the correct map, or they attempt to re-map the environment from scratch. This is the main problem: errors occur in the redrawing of their map. Due to these errors, aged rats and humans alike behave as though an environment they have visited on many previous occasions is an unfamiliar one.

Elderly humans often report similar feelings even when a spouse or friend informs them that they are in a familiar environment; this experience can be disorienting and frightening. Recall that anything that is unfamiliar will activate the amygdala (see Chapter 2) and produce a fear response. Why does the aged hippocampus fail to make accurate maps? It was once thought that the failure of the hippocampus to function normally with aging was due to massive cell loss and deterioration of neuronal connections. Today we know that the changes occurring during normal aging are far more subtle and selective. In fact, the problem with the hippocampus is due mainly to degenerative changes in a small region called the dentate gyrus.

In elderly humans and primates, the dentate gyrus of the hippocampus shows the greatest age-related loss of function. Why is the dentate gyrus so vulnerable to aging? Research published during the past 20 years suggests that this hippocampal region experiences far more age-related inflammation, and inflammation-induced pathology, than any other brain region. Inflammation that develops with normal aging is not well understood, but it may be initiated by obesity, poor diet, brain trauma, diabetes, or mutant proteins. The long-term exposure of the neurons in the dentate gyrus to brain inflammation impairs normal function and prevents the brain from making new memories or new spatial maps of the environment. Brain inflammation is also present in the earliest stages of dementia. Long-term brain inflammation greatly reduces the size of the hippocampus; this leads to serious problems with learning and individuals' ability to find their way around both new and unfamiliar environments.

How can I reduce inflammation in my brain?

Given the critical role of inflammation underlying these symptoms of dementia, it should not be surprising that long-term treatment with anti-inflammatory drugs, such as ibuprofen and aspirin, may protect against many of these age-related changes in brain health. Many epidemiological studies have discovered that daily use of high doses of nonsteroidal anti-inflammatory drugs, such as ibuprofen, for at least two years was associated with a significantly reduced risk of Alzheimer's disease. Unfortunately, anti-inflammatory drugs cannot effectively treat the symptoms of Alzheimer's disease once they have appeared nor can they reduce the extent of brain pathology associated with this disease. Furthermore, daily dosage with any of the anti-inflammatory drugs currently on the market would produce significant bleeding and discomfort in the gut.

In addition to ibuprofen, a modest amount of alcohol every day also may help to protect your brain from developing dementia. Researchers followed 3,069 people for six years and reported that people who drank one to two drinks a day were 37% less likely to develop dementia than teetotalers. It did not matter whether their drink of choice was wine, beer, or hard liquor. The reduction in risk due to regular beer consumption was similar to that associated with exercising three times a week or more. In contrast, for people 75 years of age and older with mild cognitive impairment any amount of alcohol accelerates the rate of memory decline. Alcohol provides a good example of what scientists have learned about drugs that slow brain aging: the sooner you begin the treatment, the better the outcome.

In spite of claims in the popular press or on the Internet, currently there is no treatment capable of stopping the deterioration of brain cells associated with normal aging. The most prevalent age-related disease associated with cognitive decline is Alzheimer's disease; therefore, I will focus upon what you need to know about treating this disease. For most patients, the diagnosis of Alzheimer's disease is made late in the progression of the disease. This is unfortunate because it prevents the patient from taking advantage of treatments that might slow the progression of the dementia. In addition, a better understanding of the risk factors for Alzheimer's disease might one day reduce the incidence of the disease.

So what are the major risk factors for Alzheimer's disease? A family history, particularly on the female side; the presence of specific genes; major head trauma after age 50; and depression and diabetes are all well-established risk factors. A maternal history, in particular, is related to a higher risk for developing Alzheimer's disease. A 65-year-old woman has a one-in-six chance of developing Alzheimer's disease as compared with a one-in-eleven chance for a man of the same age. In addition, the symptoms of dementia progress more quickly in women as compared with men. Finally, elderly women are more likely than elderly men to suffer long-lasting cognitive impairments after experiencing surgical anesthesia. Women also show an additional peak onset of schizophrenia at age 50 years, which is the average age for the onset of menopause in the United States. The significant changes in body and brain physiology associated with menopause appear to place women at risk of many neurological disorders.

Being well educated complicates the situation. First, the good news: two different studies of Catholic nuns have demonstrated that being college educated increases lifespan by about eight years, as compared with nuns who only finished high school. Being educated also delays the onset of symptoms of dementia associated with Alzheimer's disease. However, if you do develop Alzheimer's disease, your mental and physical decline will be accelerated, as compared with people who develop Alzheimer's disease but did not go to college. This counterintuitive finding may be due to an imbalance in neurochemistry that is beyond the coverage of this book. Next, let us turn to another age-related disease that often coexists in many patients with Alzheimer's disease, and which may be triggered by mutated genes as well as by an age-associated increase in brain inflammation: Parkinson's disease.

What is Parkinson's disease?

The symptoms of what is today known as Parkinson's disease were first described in ancient texts dating back to about 2,000 B.C.E. Much later, the physician Galen called it a "shaking palsy." Galen's depiction was maintained by Dr. James Parkinson, a practitioner of general medicine in London who is credited with the first complete description of the symptoms and progression of the disease in 1817 in a monograph titled, "Essay on the Shaking Palsy." The symptoms include the following: tremor or shaking that usually begins unilaterally in the hands and fingers when at rest; slowed movements with short shuffling steps; stiffness in the muscles of any part of the body, which can sometimes be quite painful; a stooped posture and balance problems with the typical risks associated with falls; an impaired ability to initiate or perform unconscious movements such as blinking, smiling, or swinging the arms when walking; and problems with speech, such as speaking too softly or hesitantly, or lacking emotional inflections. Dr. Parkinson identified nearly all of these signs and symptoms in the six patients who were discussed in his original essay, but he overlooked their muscle stiffness. Medical historians claim that Dr. Parkinson preferred not to touch his patients; instead, he used his long cane to poke at them during physical examinations and thus missed one of the critical diagnostic criteria for the disease.

The major risk factors for Parkinson's disease suggest the mechanisms that underlie the cause of the disease. Once again, both genes and environment play a role in this disorder. In the United States, a major risk factor is growing up in a rural environment; the cause is thought to be the early and prolonged exposure to pesticides. Other risks include exposure to heavy metals and use of an illicit opiate-like drug. A genetic risk has been identified for some families but plays little or no role for the vast majority of patients. Animal models of Parkinson's disease have provided detailed evidence on the role of brain inflammation and subsequent oxidative stress in the death of vulnerable neurons throughout the brainstem. Although it is challenging to assign specific symptoms to the death of single neurotransmitter systems, in general, the loss of dopamine neurons underlies the motor impairments, the loss of nor-epinephrine neurons underlies the lethargy and low arousal levels, the loss of serotonin neurons may lead to the presence of visual hallucinations that are often reported following the cessation of some dream sleep episodes, and the loss of basal forebrain acetylcholine neurons may underlie the symptoms of dementia. This last association with acetylcholine cell loss is important because a recent study suggested that within 10 years of diagnosis almost all patients with Parkinson's disease show some degree of dementia. Today, standard therapy involves administering drugs designed to enhance dopamine function within the synapse. Other more invasive therapies involve stimulating or lesioning specific components of the brain's motor control systems. As we learn more about how the disease develops and progresses, safe interventional therapies likely will become more effective. Next, let us turn away from pathological aging and look at normal aging—something we all hope to experience.

How does my nervous system change as I get older?

Your brain loses weight. When you were born, it weighed about 360 grams; when you were 20 years old, it weighed about 1,375 grams; and, on average, it will decline to about 1,265 grams by the time you reach 80 years of age. Most of these changes are not due to cell loss but rather to the simple dehydration of your brain and the subsequent shrinkage of neurons in vulnerable regions. Neurons lose their ability to function when they shrink. A minor contributor to the brain's decrease in size is the loss of myelin insulation around axons. The loss of myelin slows down the flow of information between neurons across the brain.

What happens to my vision?

The visual system, which provides so much essential information to your brain, shows use-dependent age-related declines in function. The eyes develop farsightedness due to the loss of lens elasticity and the subsequent flattening of the shape of the lens, reducing the ability of the lens to accommodate focusing on close objects. The increased density, as well as opacity, of the lens also increases the minimal amount of energy needed to elicit a visual response. These changes in the lens often are accelerated by long-term exposure to bright sunlight. Due to the age-associated increasing opacity of the lens, you tend to lose sensitivity to blue light; this is most noticeable to people who undergo removal of the lens and often report that blue objects are more brilliant. The retina is a thin network of neurons at the back of the eye that detects incoming light and transduces this information into electrical impulses that are transferred to the brain for processing into images. The retina is vulnerable to many of the same age-related degenerative changes that occur in the brain, such as oxidative stress and various toxins.

What happens to my hearing?

The aging auditory system develops a general loss of hearing but mostly of higher frequencies (around 3–7 KHz); sadly, the consequence of losing these frequencies is a reduced ability to perceive speech. The hearing loss is due mostly to degenerative changes occurring in an inner ear structure called the cochlea. For people living in noisy environments or using earbuds to listen to loud music, the energy of the sound is conducted into the inner ear, leading to actual cell death of sound-receptor cells.

What happens to my balance?

The vestibular system, which monitors your body's angular acceleration, as well as its sense of gravity and tilt, also resides within your inner ear. Without doubt, the most bothersome age-related problem that most people complain about is a loss of balance. This occurs in about one third of all people 40 years and older; the degeneration of vestibular nerves contributes to the high number and frequency of falls in the elderly. Dizziness is often the most common complaint that elderly patients mention to their doctors. Why? The number of neurons originating within the vestibular system in the inner ear of a 20-year-old person is about 19,000; this number declines to about 18,000 by age 60 years. Apparently, the loss of only about 5% of these neurons is sufficient to cause vertigo and problems with balance. Vertigo, a sensation that either you or the room is spinning, is a complaint of about 90% of patients seen in geriatric clinics.

What happens to the taste of food?

Your sense of taste is handled by your gustatory system. With normal aging, the absolute threshold of taste increases. Essentially, what this means is that you require more spices and more intense flavorings to enjoy your food. Your taste perception also declines with age; some familiar or favorite foods simply lose their complexity and pleasure. Part of this loss is due to degenerative changes occurring in the olfactory system within your nose. Your ability to identify unique smells peaks between age 20 and 40 years, declining slowly with normal aging. Interestingly, your ability to smell sweet and fruity odors is most vulnerable to age-related changes, while your ability to smell musky or spicy odors remains relatively stable with age.

Why do things feel differently?

Finally, your sense of touch exhibits the following changes with normal aging. More pressure must be exerted to report tactile sensation. Sensory discrimination decreases. Using a two-point threshold discrimination task, that is, how far apart do two pressure points need to be in order for you to detect two distinct points, subjects between 20 years and 36 years of age could distinguish two points 6.3 mm apart on their palms and 2.2 mm apart on their thumbs. Subjects between 63 years and 78 years of age could distinguish two points 7.8 mm apart on their palms and 3.9 mm apart on their thumbs. Clearly, just about every aspect of your aging nervous system changes with age—which raises the next most important question.

What can you do to slow aging?

Stop eating so much! Especially avoid red meat and dairy products. There are many other healthier dietary sources for iron and calcium. The benefits are now well known: dietary restriction slows brain aging and protects against many neurodegenerative diseases. The exact biological mechanisms underlying the beneficial effects of dietary restriction are not completely understood, even though the effects have been observed in many species, ranging from worms to monkeys. As already discussed, the current evidence suggests that a reduction in free radical formation and subsequent inflammatory processes, as well as alterations in the expression of genes that regulate our circadian rhythm, underlies at least part of dietary restriction's beneficial effects.

How much should you restrict your diet?

Not that much. In a recent study, a large group of monkeys, ranging in age from middle-aged adults to the quite elderly, were fed 70% of their free-feeding diet for about 15 years. This study is valuable because it investigated the effects of a reasonable reduction in total calorie intake and the monkeys were on this restricted diet for a significant portion of their lives. Essentially, for a normal human male eating 2,000 calories per day, this would be about 600 fewer calories per day. For comparison, 600 calories would be about one cup of roasted almonds, or a typical 100 gram bar of dark chocolate, or a typical Cold Stone Creamery dessert (with the M & M's of course!). Obviously, the monkeys were not on a starvation diet.

Because they consumed 30% fewer calories, the brains and bodies of the monkeys on the restricted diet aged significantly more slowly. Although several brain regions showed benefits, those brain regions that evolved most recently, such as the frontal lobes, which also tend to be more vulnerable to the consequences of aging, showed the greatest beneficial response to the dietary restriction. Monkeys on the restricted diet developed fewer age-related diseases, had no indication of diabetes, exhibited almost no age-related muscle atrophy, and lived much longer than their free-feeding compatriots.

Take note of the crucial fact that these monkeys did not exercise their weight off; they simply consumed fewer calories. Exercising is never going to be as beneficial to your brain and body as restricting the number of calories you consume! By now you can probably guess why—exercising involves lots and lots of breathing and encourages you to consume lots of energy.

What is the consequence of being overweight as I get older?

Obesity leads to brain shrinkage and increases your risk of developing dementia. One recent study demonstrated that being obese at midlife is a strong predictor of dementia in later life. The obese elderly also have more impaired learning and memory abilities than thin elderly people. Patients with Alzheimer's disease who are obese develop more pathology in their brains and demonstrate a very rapid decline in mental function. How does obesity contribute to brain shrinkage and dementia? Fat cells in the body produce inflammation by releasing specialized proteins called cytokines. The more fat cells you have, the more cytokines are released into your blood every hour of every day. Cytokines induce shrinkage of brain regions that are required for making new memories and for recalling old ones. If the obesity-induced inflammation lasts for many decades, there is more brain shrinkage and greater memory loss.

Fortunately, the sooner one loses the body fat, the sooner the brain can begin to recover. The true culprit is body fat. Older people who have relatively more visceral fat than subcutaneous fat, and thus might appear to be thin, are also at increased risk for diabetes, metabolic syndrome, and increased mortality. What would happen if these harmful fat cells were simply removed? Exercise can shrink fat cells, but only liposuction can remove them from the body. A group of scientists investigated this question by conducting three very clever experiments on obese and normal-weight mice. Mice demonstrate an identical vulnerability to the negative consequences of body fat on brain health as humans. First, a group of obese mice exercised on a treadmill. As expected, these mice reduced belly fat, reduced the level of inflammation in their body, and significantly restructured how their brains functioned at the cellular level, leading to greatly improved memory.

In the second study, the scientists surgically removed abdominal fat pads from a similar group of obese mice; that is, the mice underwent a standard liposuction procedure. The results were identical to those produced by running on the treadmill: inflammation was reduced and the mice became significantly smarter. These findings confirm many recent studies that have documented the ability of fat cells to impair brain function and accelerate aging.

In the third study, the scientists did something truly astonishing; they transplanted fat pads into normal, healthy-weight mice. The impact of the additional fat cells was immediately obvious: the mice showed increased signs of brain and body inflammation, and they developed harmful changes in brain structure and function that led to a significant memory impairment.

Today, an overwhelming amount of scientific evidence across a wide spectrum of medical disciplines strongly argues that obesity accelerates the aging process, impairs overall cognitive function and, ultimately, is responsible for numerous metabolic processes that ultimately kill us.

What can you do about your aging brain?

As you have just discovered, medical research has found some great answers. To save you the time it would take to search through a diverse collection of epidemiological studies (some of which have already been discussed), here is a compilation of the best scientifically defendable advice for living a long life and having a healthy brain: be female, drink lots of coffee, choose your parents carefully, eat as little as possible, breathe as little as possible, move as little as possible, be born in May, be tall and have a large head, develop arthritis so that you can justify taking lots of anti-inflammatories, and drink moderate amounts of beer every day. If you must eat, then only dine during the early part of your daily biorhythm.

Do miracle cures for brain aging exist?

In the void of what we have yet to discover about brain aging lie numerous unanswered questions and unproven theories. Unfortunately, countless myths have been invented to fill this void of ignorance; among these myths are those concerning age- or disease-related mental decline and the benefits of alternative, nonscientific remedies purported to restore, or even enhance, brain function. A truly remarkable variety of medical interventions have been concocted out of uneducated interpretations of genuine scientific facts and promoted by people who become victims of their own wishful thinking. We have all encountered these people; they are sincere but appear deluded and fixated on eccentric theories—and always, without fail, they are confident that their knowledge will revolutionize science or society if only somebody would just listen to them. We ourselves are vulnerable to these misconceptions because we desperately want them to be true.

Inevitably, we become dreadfully disappointed when our expectations are not met. Maybe the problem was you, maybe you were not sufficiently persuaded by all of those wonderful claims by your friends on the Internet. As Tinker Bell said, "You just have to believe!" Shouldn't your good thoughts and a positive attitude be sufficient? No. Results of many carefully controlled studies have failed to show that psychological interventions, such as simply being happier, practicing positive thinking, or receiving the prayers of devout friends, have any impact on the survival of cancer patients.

Today, the Internet, magazines, and slick television infomercials are using sloppy interpretations of questionable scientific studies to mislead desperate people. Usually, the only consequence for exploited victims is the loss of money or a delay in seeking treatment that might offer some true medical benefits.

Our brains change throughout our lives, and not always for the better. Why do they change? There are many causes of cognitive decline, including the long-term consumption of some drugs—both licit and illicit, dementia and various diseases of the brain and body, head injury, hormone imbalance, dietary nutrient deficiency or excess, heavy-metal toxicity, sleep deprivation, and prolonged stress, to name only a few. The treatments are as varied as the causes. Some modern treatments for many of these ailments are relatively effective for many people. In contrast, no treatments currently available can reverse one of the biggest causes of cognitive decline: normal aging.

Once again, the failure of modern science to provide an effective intervention has left a void of ignorance. This has been filled by con artists, who offer usually harmless products for sale and claim that these enhance brain function as we age. The most distinctive feature of any useless elixir is that it is always 100% effective. In contrast, no scientifically tested drug would or could ever make such a claim. In general, most of these unproven elixirs contain common stimulants such as caffeine or sugar in order to enhance one's level of arousal. Unfortunately, stimulants only enhance performance, not true intelligence. The classic brain stimulants already discussed—coffee or nicotine—might improve performance, engaging certain neurotransmitters in the process, but they do not raise one's IQ score, and they do not stop normal age-related cognitive decline.

Thus far, however, no one has been able to design a drug therapy that can make a person smarter in any significant way. If you look at the so-called memory boosters and cognitive enhancers on the market today, you will find that they contain caffeine and sugar and some peculiar amino acids and a few vitamins that together do nothing except make you a little poorer. At this point in time in the 21st century, nothing—let me repeat that, nothing—can truly make you smarter; thus, do not waste your money on any product that promises to do so.

One thing, however, is certain: someone, somewhere is now selling "the cure" for mental decline. Everyone would prefer to defy the aging process by simply taking a pill and being able to eat with impunity everything we desire rather than following the standard prosaic advice about moderate, healthy eating. The fact that science has not yet invented a true brain enhancer has not stopped people from selling drugs, ancient elixirs, unusual therapies with mystical names, and hundreds of books that all boast of the properties of this or that miraculous, age-defying brain booster. If someone stands to gain financially from your gullibility, then what he or she is selling is probably useless, and there is no guarantee that it is safe. Any cursory search on the web brings up countless potions containing useless, scientifically discredited but rather harmless ingredients.

Why do so many people fall under the spell of charlatans?

Why do people believe that these magical elixirs are effective for them? The answer is easily summarized in three little words—*the placebo effect.* Essentially, we very badly want these elixirs to do something, anything, to slow down the aging of our brain; so we fool ourselves into thinking that they do. After all, you have just spent a lot of money on this pill! Placebos, of course, must be expensive. No one would ever believe in the effectiveness of an inexpensive magical elixir; after all, they are special and, therefore, must be expensive. When challenged, spokespersons for these fraudulent products often claim that medical science has ignored their wonderful product because doctors simply do not want you to be healthy. Not true. Scientists have spent years testing many of these compounds; their conclusion is that these products are as useless as they are expensive. Fortunately, most of them are so utterly inactive that they will not harm you. One of the best examples of a useless antiaging product is an extract from the Ginkgo biloba plant.

What about Ginkgo?

The first challenge in using Ginkgo or any other plant product is knowing how much to use and which component of the extract is most effective. When ancient Chinese herbalists recommended that their patients take Ginkgo biloba, or any number of other plant extracts that have been prescribed during the last two millennia, they always estimated dosage based on past experience. But plants are complicated organisms that produce a large variety of molecules, some of which are active in the brain, some of which are not active in the brain but are quite nutritious, and some of which are simply inactive. Therefore, how much of any particular extract should be taken by a person who seeks the benefit that Ginkgo might offer? No one knows! The studies necessary to establish a truly effective dose are exceptionally expensive and, therefore, have never been performed rigorously. In order to avoid such expensive testing, the manufacturers, with the help of politicians, have had their products designated as nutritional supplements rather than as drugs.

Studies conducted on these nutritional supplements are often poorly designed and have various methodological problems, such as inadequate sample size (the number of subjects in the study) and lack of a double-blind, placebo-controlled paradigm, the gold standard of modern scientific research. This paradigm means that no one involved in a drug trial—including its investigators and its subjects—knows which tested substance, whether an active drug or a placebo (usually an inactive form of the drug under study or a sugar pill), is being administered. The purpose of this approach has to do, again, with bias: to keep investigator and subject bias from influencing the trial's results. In fact, on the rare occasion that this standard has been applied to studies on alternative medicines such as Gingko biloba, the results have been negative. For example, a pair of very large clinical trials that followed the health of more than 3,000 people of various ages for eight years clearly demonstrated that Gingko biloba cannot influence the development of age-related memory problems. Another trial indicated that the use of Gingko may actually be harmful and increase an individual's risk of stroke (i.e., when a blood vessel in the brain becomes blocked and shuts off blood flow). These are, however, just a handful of studies, and much more high-quality research needs to occur before the effectiveness of Gingko biloba and other herbal products is irrefutably proven or disproven.

In the meantime, most manufacturers of these products prefer to err on the side of selling diluted samples, thereby avoiding any toxic side effects and potential lawsuits from people who survive the experience. But this is still no guarantee that the samples are safe. Unacceptably high levels of pesticides and carcinogens have, for example, been found in a large percentage of imported samples of Gingko biloba and many other herbal medications. These concerns aside, many people are convinced that they benefit from substances like Gingko biloba or the countless other useless products on the market, such as extracts of deer antlers or sea horses, that promise enhanced cognitive function. Why? Because they want these drugs to do something and, therefore, fool themselves into thinking that they do. We all are subject to this faulty thinking from time to time.

What is pseudoscience?

In addition to drugs and herbals that are marketed using pseudoscientific logic, there are almost as many nondrug interventions for your brain that also lack any shred of scientific proof. These interventions usually invoke the actions of some mystical force that physicists have yet to discover. The fact that these interventions lack any scientific support does not deter desperate people from seeking them out and, most important, paying for them. Craniosacral therapy, ear candling, magnet therapy, crystal healing, cupping, Rolfing, neurolinguistic programming, psychokinesis, and primal therapy are just a few of the frequently mentioned examples of completely ineffective interventions. In addition, "energy medicines," which involve "laying on of the hands," or various types of hand waving above the body, or any of the numerous naturopathic practices, provide no medical relief beyond the placebo effect.

What is the placebo effect?

Much has been written about the value of the placebo effect in the practice of medicine, but how this effect emerges and whether it can be controlled are issues that are not yet understood. Essentially, scientists have analyzed the effect based on results of placebo-controlled studies of actual drugs or have compared only the effects of a placebo against the consequences of no treatment at all. Their findings have been intriguing, if still largely inconclusive. However, in one area of study that is not directly related to an actual treatment, the findings are more definitive. Numerous meta-analyses (which are combined analyses of other researchers' data) have shown that the perception of pain can be statistically demonstrated to be influenced by our minds. Scientists refer to this as the emergent property of our brains. This influence of our thoughts and expectations on how we experience pain is a true placebo effect.

Perhaps one particular study has shown us the location of the placebo effect. Scientists measured pain perception in two groups of people, devout practicing Catholics and professed atheists and agnostics, while they viewed an image of the Virgin Mary or the painting of Lady with an Ermine, by Leonardo da Vinci. The devout Catholics perceived electrical pulses to their hand as being less painful when they looked at the Virgin Mary than when they looked at the da Vinci work. In contrast, the atheists and agnostics derived no pain relief while viewing either picture. Magnetic resonance imaging (MRI) scans demonstrated that the Catholics' pain relief was associated with greatly increased brain activity in their right bottom-lateral frontal cortex. This brain region is believed to be involved in controlling our emotional response to sensory stimuli, such as pain.

Other studies using brain imaging techniques to show correlations between brain activity and the extent of reported placebo effects have demonstrated that some people show greater placebo responses than others, but that everyone appears to be capable of having such a response. There is also increasing proof that the use of placebos might benefit people with Parkinson's disease, depression, and anxiety. In the future, with better testing measures, scientists likely will demonstrate how the placebo effect influences many aspects of our health. In short, the placebo effect is real; we simply do not understand entirely how it works, but the evidence thus far is truly remarkable, particularly with regard to pain. Some people are able to block incoming pain signals or alter how they are perceived. And without a doubt, your mind can make the experience of pain more or less agonizing depending on how you feel—for example, if you are fatigued, anxious, fearful, or bored, then pain becomes more intolerable. Recent studies have identified specific sets of genes, referred to as the "Placebome" that may contribute to the placebo effect. In the future, a detailed analysis of your genetic makeup may allow your doctor to predict clinical outcomes better and potentially allow the judicial use of "effective" placebos.

Although we do not yet know how the placebo effect works in the brain to influence this process, we do know that it comes into play, and sometimes in surprising ways. For example, the color of the pill you take influences your expectation of what it will do to you. Obviously, pills can be made any color, yet most people like their antianxiety pills to be blue or pink or some other soft, warm color; they prefer their powerful anticancer pills to be red or brightly colored. Americans do not like black or brown pills, in contrast to the preference of people in the United Kingdom or Europe. Thus, almost everything that Americans buy over the counter is a small white, round pill. Yet big pills, or pills with odd shapes, also are assumed to be more powerful, or just simply better, than tiny round pills. Sometimes, a simple change in color or shape restores a drug's ability to produce a placebo effect. And sometimes, the effect comes from the pill-taking regimen. For example, you expect that when you are instructed to take a medication only during a full moon, or only every other Thursday, it must be extremely, almost mystically, effective. Herbalists often take advantage of this concept by recommending odd or excessive dosages of peculiar-looking pills or foul-smelling potions. We all want to believe that the pills we take will help us feel and function better; fortunately, thanks to the poorly understood phenomenon of the placebo effect, we do sometimes, but only for a while, benefit even from the most bogus of potions and pills. If all you are getting is a sugar pill, then does it really matter whether you are fooled into believing the lie? Possibly; it depends on the cost of the sugar pills and the risk one assumes by not taking a medicine of proven effectiveness in a timely fashion for a medical condition. Finally, nothing has ever been proven to enhance brain function; only caloric restriction can slow brain aging. Just think about how much money you are going to save by not wasting it on unproven alternative therapies and by consuming much less food.

6

HOW DOES MY BRAIN
ACCOMPLISH SO MUCH?

I originally considered beginning this book in the traditional style with a discussion about the basics of brain anatomy. I decided not to follow that tradition because I wanted you to become excited about what your brain can do before investigating the precise anatomical structures that achieve those amazing feats of learning and sleeping and love.

Let us start our exploration of basic brain science by drilling into the top of your head! The tip of the drill has traveled only about one half inch through your scalp and skull, and we immediately encounter three layers of protective membranes called meninges; their names are the dura (the outermost layer), arachnoid (the middle layer), and pia mater (the innermost layer). A person has meningitis when these membranes become infected. Between the arachnoid and pia layers is a space filled with a clear, colorless liquid. This liquid is cerebrospinal fluid and is essentially blood that has been filtered of cells and most proteins. Freshly produced cerebrospinal fluid is constantly rinsing your brain. The amount of fluid produced is impressive: every ounce of cerebrospinal fluid is completely replaced about four times every day. If the constant flow of cerebrospinal fluid is impeded in any way, the fluid will quickly accumulate inside your skull, thereby increasing intracranial pressure, pressing the brain against the inside of the skull, squeezing the small blood vessels that feed the brain until they close, and ultimately leading to the death of brain tissue. This condition is called hydrocephalus; it occurs more often in infants than older adults and can be fatal if not corrected immediately. Your skull has a fixed volume; this fact often places your brain in serious peril.

The brain is submerged in an ocean of cerebrospinal fluid inside your skull. Why go to all of this trouble to keep the brain afloat? The answer is related to the fact that your big brain weighs a lot: about three pounds. However, when floating in the cerebrospinal fluid, the net weight of your brain is equivalent to a mass of only 25 grams—that is less than an ounce of beer. How is this possible? If you have ever experienced floating in the ocean, the salt water made you buoyant and you floated easily on the surface with very little effort, as though you weighed much less. The salty cerebrospinal fluid provides the same benefit; it allows the brain to maintain its density and shape without being crushed by its own weight. This same principle of buoyancy allows seagoing mammals, such as whales, to become very large; however, once they have left the salt-water ocean, they quickly succumb to the consequences of their own weight. If your brain were resting on a table, its own weight would quickly crush the small blood vessels supplying it, killing the cells on the bottom of the brain.

In addition, by floating within the skull, bathed in cerebrospinal fluid, the brain is protected from injury when the head is jolted around quickly. Unfortunately, when the head is violently displaced due to a car accident or a blow to the head associated with playing football, soccer, or field hockey, this fluid buffer cannot prevent the brain from colliding with the inner surface of the skull. If the brain does collide against the skull, the outer layer of the brain, the cortex, can become attached by scar tissue to the inner surface of the skull, leading to the death of outer layers of the cortex. Repeated violent blows to the head, such as those experienced by boxers and other athletes with a history of repetitive brain trauma, ultimately leads to large sections of the cortex becoming stuck to the inner surface of the skull. If the injury to the cortex continues and becomes widespread, the resulting loss in brain function is called dementia pugilistica; in the 1920s these symptoms were called the punch-drunk syndrome. Today, due to the notoriety professional athletes have drawn to this condition, this progressive degenerative disease is referred to as chronic traumatic encephalopathy. The damage to the cortex associated with this injury may spread to involve nearby subcortical brain areas that are responsible for movement and sensory processing, producing symptoms such as stiffness, slowness, and walking or balance problems. When this happens, the result is Parkinson's pugilistica. The boxer Muhammad Ali demonstrated many of the symptoms of both dementia pugilistica and Parkinson's pugilistica.

As we continue to drill into your head, the next thing we encounter is your cortex. The cortex is a thin sheet of cells that covers the brain; it is responsible for the extraordinary abilities that you depend upon for normal everyday life, such as thinking, feeling, seeing, hearing, and touch. Given the importance of the cortex for so many abilities that truly define our species, this chapter will focus primarily upon its structure and function. If you imagine yourself miniaturized and standing on the surface of your brain, the cortical surface appears like a series of rolling hills, which are called gyri, and valleys, which are called sulci. Anatomists have divided the brain into four separate lobes according to the location of specific and very deep sulci. The lobes were given the names frontal, temporal, occipital, and parietal (derived from the names of the skull bones that overlie the lobes). Overall, your brain, as well as the brain of your cat and the mouse it is waiting to catch, is organized so that the back half receives incoming sensory information and then processes it into your own very personal experience of the here and now. The front half of your brain, the frontal lobes, is responsible for planning your movements, usually in response to some important incoming sensory stimulus, such as someone's voice telling you that it is time for dinner. The lobes in the back half of your brain process the voice that you hear, smell the aroma of food cooking, feel a craving for food as your blood sugar levels fall, and sense that it is late in the day and the sun is setting and the room is getting darker; thus, it must be dinnertime. This information is funneled into the front of your brain, which then makes a decision to move the front end of your feeding tube toward the smell and the voice to obtain a reward—food and survival for another day! Your brain has successfully performed its most important tasks for the day, and it now can rest with you in front of the television.

How does your brain manage all of this every day? Neurons—billions of them. Your brain, as well as the brain of your dog and the flea living on its back, is composed mostly of cells called neurons that interact both electrically and chemically with one another. Your brain also contains some supporting cells called glia. The average adult human brain contains about 90 billion neurons, give or take. A typical neuron possesses a cell body, dendrites, and a long single axon. Dendrites are thin filaments that extend away from the cell body, often branching multiple times, giving rise to a complex tree of dendritic branches. An axon is a special narrow extension that projects away from the cell body and travels for a rather long distance. Some axons can extend for almost three feet in humans. The cell body of a neuron frequently gives rise to multiple dendrites, but never to more than one axon. An individual neuron will communicate via its dendrites and axon with about 7,000 other neurons. The connections are far more complicated than those of your local telephone network.

If you were to scoop out a very small section of cortex, you would find it packed with neurons, glia, blood vessels, and very little else. Most of the space between neurons is filled with astrocytes, a type of glia. Astrocytes have another more notorious role; astrocyte tumors make up the majority of brain tumors. Astrocyte tumors are categorized as either low-grade, which are usually localized and grow slowly, or high-grade, which grow rapidly. Most astrocyte tumors in children are low grade; in adults, the majority are, unfortunately, high grade. Tumors that consist of only neurons are quite uncommon and, when they do occur, tend to be less aggressive. The contrast between these two types of tumors, glial versus neuronal, is likely related to the fact that the normal function of glia is to multiply in number following certain types of injury or infection, while neurons are genetically programmed to never multiply once they have achieved adult status.

The brain is densely packed with blood vessels. Neurons are never more than a few millimeters away from one because a constant supply of oxygen-rich blood is critical for normal brain function. Blood flow to the brain, about three cups per minute, is carefully regulated because too much or too little blood flow can be harmful to the brain. Normal brain function is rapidly lost if the supply of oxygen is interrupted for even a few moments. Sometimes, the temporary loss of blood supply to the brain can produce quite bizarre experiences, such as near-death visits to heaven.

Are near-death visits to heaven real?

No, they most certainly are not. Unless you want them to be. Near-death visits to the spiritual realm usually begin in a hospital emergency or operating room and are associated with the failure of normal heart function due to trauma, extended seizures, or cardiac arrest. When blood flow to the brain slows down, the supply of oxygen to the billions of individual neurons falls quickly. When the supply of oxygen drops, even just a little, 60% of the brain's dopamine is very quickly converted into an entirely different molecule called 3-MT. Until a few years ago most textbooks stated that 3-MT was completely inactive in the brain. Our understanding of the actions of 3-MT in the brain improved when neuroscientists discovered that 3-MT acts in the same way as many hallucinogens, such as LSD and ecstasy. Consider this scenario: just as the blood flow to a patient's brain is decreasing, the brain is spontaneously producing very high levels of a powerful hallucinogen. What might that feel like? People who have survived such near-death experiences often report floating through a blissful spiritual world that is full of love. People who have used LSD and ecstasy report a very similar emotional and sensory experience. Thus, the spiritual, pleasant, loving near-death experiences due to elevated levels of 3-MT are probably a consequence of the reduced blood flow to the brain. The mythical hallucinations associated with near-death experiences, and the ease with which we believe them to be true, demonstrate how vulnerable our sense of reality is to imbalances in brain chemistry.

How do nutrients and drugs get into my brain?

Astrocytes carefully control what is able to cross from blood to brain. Astrocytes are a critical component of your "blood–brain barrier." The blood–brain barrier permits the easy entry of only a few substances into the brain. Fat-soluble substances can enter the brain easily. Very small molecules, particularly if they do not carry an electrical charge, usually get through the blood–brain barrier. The brain actively imports the nutrients it requires from your diet through the blood–brain barrier. Some regions of the brain lack any blood–brain barrier. The barrier does not form in these brain regions so that your brain can monitor levels of specific chemicals, such as the presence of sugar in the blood. As you will see later, the brain is only capable of sensing the presence of sugar, actually glucose, in the blood; it cannot sense the levels of proteins or fats in the blood. The ways in which the brain controls eating behaviors will be discussed later.

The blood–brain barrier is often the Achilles heel of so many drugs that might offer something beneficial to the brain—the drugs never get into the brain. Your brain rests comfortably behind a biological firewall called the blood–brain barrier.

Every day you consume chemicals that would produce significant changes in brain function—if they could get across this barrier. There are times, of course, when you might prefer that drugs could get across this barrier more easily. Only about 5% of all of the drugs currently available by prescription can cross the blood–brain barrier. Today, drugs that are designed to treat disorders of brain function are specifically designed to cross this barrier. Obviously, if a drug cannot enter the brain, it is going to have a difficult time influencing brain function. One recent case illustrates this point. Prevagen is a commercial product currently being aggressively marketed as a memory supplement; its main ingredient is apoaequorin, an incredibly large, highly water-soluble molecule that is an excellent example of a molecule that cannot cross the blood–brain barrier. Any benefit it provides to memory is, therefore, due to the placebo effect. In addition, the U.S. Food and Drug Administration (FDA) recently issued a warning that Prevagen may produce serious side effects that are not due to the placebo effect.

How is the cortex organized?

The neurons within your cortex are organized into columns of cells that are arranged in six layers. Each column of cells is oriented perpendicularly to the cortical surface and consists of about 100 neurons. The cells in each column are highly interconnected with one another. Animals with bigger brains simply have many more of these columns of cells. Think of the cortex as a bed sheet, with a thickness of about three to four millimeters for most mammals; in contrast, however, the width and length of the bed sheet vary from twin size (dogs) to queen size (humans) to king size (whales). The evolution of the cortex led to bigger bed sheets that vary only a little in thickness. The increase in size was accompanied by a buckling and rippling leading to the formation of a cortical landscape characterized by small hills and valleys; this folding process made it possible to fit a large cortical sheet within a small skull. Overall, the larger and thinner the cortex, the more folded it is. For example, the large brains of dolphins and whales have exceptionally thin cerebral cortices, which are remarkably folded with lots of small gyri. Across the many different species that have been studied, the folding pattern has remained surprisingly consistent; for example, motor functions are always up front, while sensory abilities are located in the back half of the brain.

Why aren't human brains bigger?

Neuroscientists have wondered why we never evolved bigger, and perhaps smarter, brains. Why isn't the cortex thicker? Why didn't the forces of evolution give us more columns and make a bigger sheet of cortex? Either of those two advances might have allowed humans to be much smarter. The reason the cortex is not thicker is partly because there would not be enough room for all of the axons necessary to connect all of those extra neurons to one another. Fancy wiring has made us a very successful species; however, not having additional room for all of those extra wires (i.e., axons) has limited humans from becoming even smarter. Imagine the brain as a pair of football-shaped hemispheres that are covered by a thin layer of cortex: so what fills the inside of the football? Most of the interior of the brain is filled with the axons (i.e., the wires) that are necessary to connect all of your neurons. Just for a moment, let us assume that we could add another layer to the surface of the cortex. If this happened, it would become necessary to greatly increase the size of the entire brain in order to make room for all of the extra wiring required to allow the additional layer of neurons to communicate with all of your other neurons. That is the problem: the additional wiring might make us smarter, but the wires would require a lot more room inside the brain. This leads to a much bigger problem: bigger brains require bigger skulls; bigger skulls require much bigger birth canals; bigger birth canals are just not an option for a bipedal, vertically symmetrical animal living on dry land, such as a human. Overall, it appears as though the human brain has evolved to be as big as possible, given the anatomical constraints imposed by the fact that it resides inside your skull, which spends time developing inside your mother.

How does the cortex develop?

Imagine that you are watching as an embryo begins to grow inside the womb. Focus on just the front of the embryo where the brain is beginning to take shape. At this stage neurons begin their lives as protoneurons resting in the wall of a fluid-filled neural tube. Each protoneuron then slowly crawls outward over the top of its neighbors until it reaches its final resting place. In this fashion, each of the six layers of your cortex is deposited like a layer cake, starting with the lowest layer at the wall of the neural tube; then, progressively newer neurons-to-be crawl over their siblings to deposit themselves on top as a new layer. Once deposited, each protoneuron differentiates into a mature adult neuron and never undergoes another cell division. The ultimate size of a brain is determined by how many cell divisions occur within the walls of the neural tube before the cells start climbing.

Differences in brain size between most species are largely due to variations in the size of the cortex. This discussion, therefore, is focused on the cortex. Recent research has discovered that in the mouse brain these primordial neurons undergo nine cycles of division, taking about five hours to complete the process. In the monkey brain, these primordial neurons undergo 18 cycles of division and migration, taking about 20 hours to complete the process. In humans, our primordial neurons undergo 20 cycles of division and migration, taking about 22.5 hours to complete. It is amazing how little time it takes to build the basic framework of the cortex. The center of higher mental functions for humans, the thin layer of cortex on the surface of the brain contains at least 100 billion neurons, and 100 million meters of axons that connect these neurons to one another.

Scientists now know that the size difference between the brains of *Australopithecus* and *Homo erectus* was due to only one additional cycle of cell division—just one! Think of the simplicity of this evolutionary process; the way to build a bigger brain is simply to wait and do nothing for a couple of hours, giving the protoneurons a chance to undergo just one more cell division before beginning their differentiation into adult neurons.

In summary, the evolutionary solution to making a brain with more capabilities was to generate only six layers of cells in the cortex but expand its total surface area. A recent study of many different mammalian species discovered that cortical thickness varies across species by only a few millimeters; of course, there are some interesting exceptions. Manatees and humans both have unusually thick cerebral cortices on average. Does this mean that manatees are as smart as humans? Not at all! Remember, the most important feature underlying intelligence is fancy wiring between neurons, not the size of the brain. Bigger brains do not guarantee greater intelligence.

How do male and female brains differ?

Male and female human brains have many anatomical differences—for example, the corpus callosum is thicker in females than in males. This difference is evident even as early as the fetal stages of development. Is this an important difference? The corpus callosum is a large bundle of axon fibers that allows the two hemispheres of the brain to talk to each other. Having a thicker corpus collosum would offer the opportunity for increased cross-talk between brain hemispheres; this additional avenue of communication is thought to underlie better language skills in females and also may underlie the fact that boys have more learning disabilities and dyslexia than do girls.

Overall, probably related to their thicker corpus callosum, female brains excel at the intercommunication between hemispheres while male brains show more connectivity within each hemisphere. You will recall that the back of the brain is involved in sensory perception and processing while the front half of the brain controls coordinated movements; the greater intrahemispheric connectivity in male brains may explain why some men are generally better in activities that require skilled eye-hand coordination.

Are bigger brains always better?

The size of the brain correlates with longevity in mammals. Larger brains have an initially slower growth period and then spend a longer time giving birth to new neurons. Smaller brains have a faster initial growth period and then neuronal development terminates earlier in the development of the organism. Consequently, human babies are born with a brain encased within a soft skull that is just small enough to squeeze through the birth canal and then quickly begins to grow much bigger. Unfortunately, this means that human babies are born quite vulnerable and are not able to thrive without significant and prolonged parental intervention.

Having a big brain is never enough. The African elephant brain has three times as many neurons as the human brain. However, the African elephant has far fewer neurons where they are truly needed—in the cortex. If the African elephant brain had evolved a few critical modifications, African elephants would be ruling the world and humans would be performing circus tricks for their benefit.

Chimpanzees and humans begin their lives with about the same number of neurons per cubic inch of cortex, but as the human brain develops, it systematically destroys neurons that get in the way of the growth of axons that are connecting the various parts of the brain to one another. It is astonishing to realize that we humans became smarter, and more intellectually agile, than our closest evolutionary relative by simply killing off excess neurons to obtain the benefits provided by more complex wiring between brain regions. The frontal lobe of the human brain performs such astonishing mental exercises because its wiring is much more complicated than what is seen in the temporal or occipital lobes. So what is the take-home message? It would be that when it comes to brains, the complexity of the wiring matters; overall size is generally less important.

What is neurogenesis?

It was once believed that all of your neurons were born while you were developing inside the uterus and that only a small percentage were added during the first few years of life. The assumption always had been that adult brains possessed all of the neurons that anyone would ever need. Apparently, this is not true; we actually require a constant supply of new neurons when we are adults, too. Human brains give birth to approximately 1,400 new neurons every day. This process of neuronal birth is called neurogenesis and only occurs within the hippocampus (in humans), a region of the brain within the temporal lobe that is critical for learning and memory. Today, partly due to the above-ground detonations of nuclear bombs that were conducted between 1945 and 1963, we know quite a lot about the birth of new neurons in adult brains. The atmospheric radioactive carbon atoms that were released by those explosions have been incorporating themselves into the DNA of our dividing neurons, providing a time-stamp of when each neuron was born. What do we do with all of those new neurons every day? Their ultimate fate is unknown: some of them are most likely being used by the hippocampus to help incorporate our daily memories.

How do we think so fast?

Neurons communicate with one another via their axons. Axons are conduits for an electrical disturbance called an action potential. Action potentials begin within the cell body of a neuron; once this electrical disturbance is released from the cell body, it ripples its way along the length of the axon until it reaches the end of the axon, where it induces the release of chemicals. The speed of the action potential along the axon, up to about 120 meters/second (or about 3,000 miles per hour), is directly related to two independent factors—the diameter of the axon and the presence of a thick insulating layer of tissue, called a myelin sheath, that is wrapped around most axons. Very thin axons are the slowest; thicker axons conduct action potentials much faster. Some species, such as the intelligent octopus, evolved very big axons and large heads relative to their bodies. This works if you are a mollusk living in the salt water ocean. Our small skulls do not provide enough room for these large axons to live. A myelin sheath is analogous to the insulation on electrical wires; it prevents electrical signals on one axon from leaping over to nearby axons. In summary, during the evolution of bigger brains, myelin sheaths allowed axons to remain thinner but maintain a faster conduction speed for the action potential, thus providing faster information processing. All of these abilities developed within a brain small enough to pass through the birth canal.

What is multiple sclerosis?

Considering everything that myelin does for brain function, it is probably not surprising that your brain cannot function normally without myelin sheaths wrapped around many of its axons. To fully appreciate why, just imagine the consequences of losing some or all of the insulation on the wires inside your computer, or inside the walls of your home; nothing would work correctly. For example, you might throw a switch hoping to send electricity to your television; the electrical signal might randomly jump to another wire and proceed to a completely unexpected location, such as your toaster. Unfortunately, this misdirection of electrical signals occurs in the brains of people with multiple sclerosis. The immune system of these patients attacks the myelin sheath surrounding axons in their brain and spinal cord. Multiple sclerosis usually begins between the ages of 20 and 50 and is twice as common in women as in men. As the disease progresses, the passage of the action potential along axons becomes slower and slower until sometimes the action potential never makes it to its intended destination. Visual, motor, and sensory problems are the most common reported symptoms. Over time, as more myelin is lost, patients with multiple sclerosis develop problems controlling their thoughts and emotions. Interestingly, the symptoms worsen when the body gets overheated due to exercise, fever, or hot weather; this is likely due to the effect of the heat on the nerve's ability to conduct an action potential. Clearly, the failure of action potentials to reach their intended destination interferes with normal brain function and makes the lives of these patients quite challenging.

Does my brain work perfectly all of the time?

Textbooks often present the brain as working perfectly at each and every step along the way to consciousness and normal thought; in fact, nothing could be further from the truth. Even if the action potential does successfully complete its journey to the end of the axon, there is no guarantee that anything will happen. There is considerable failure to function and outright chaos associated with much of what happens in your brain. Yet, somehow we do achieve consciousness in spite of the inherent errors associated with the challenge of getting all of the right processes to occur at precisely the right times. For example, under normal circumstances the arrival of the action potential at the end of the axon will induce the release of specific chemicals. Sometimes, nothing is released or too few of these chemicals are released. The chemicals released from the ends of neurons are called neurotransmitters; they instruct the next neuron to initiate its own action potential, and so on.

This complex and surprisingly slow electrical-to-chemical communication process is happening in your brain while you are reading this sentence: lots of electrical ripples propagated along your visual axons inducing the release of neurotransmitter chemicals in the part of your brain that handles vision. This visual brain region then sent electrical action potentials off into the vast reaches of your brain to inform those brain regions that you are reading words expressed as black lines on white paper, and that these words have meanings that your brain must interpret. The arrival of action potential rippling along an axon is supposed to induce the release of neurotransmitter chemicals onto the next neuron, and so on and so on. This is how information and thoughts are processed. Various aspects of this process are happening, completely or partially, all over your brain, all of the time, and are the basis of your consciousness and everything you will ever experience. The take-home point is that our brain does not function perfectly all of the time; however, ultimately, we manage to endure with what we have evolved.

What are neurotransmitters and what do they do for me?

Within the brain, most of the major structures evolved as small clumps of cells, called nuclei or ganglia, which are involved in related functions. Some ganglia control movement, some control body temperature, and some control your mood. Overall, the basic plan, whether you are an octopus or a human, is that neurons communicate with one another in order to facilitate the sensing of the external world and the internal events taking place inside the body. The brain then decides which behavior to elicit in order to improve its chances of survival and propagation of its species. For humans and some other animals, there is sometimes a more ephemeral goal: Do something that brings pleasure. For humans during the 1960s, this was succinctly re-envisioned as sex, drugs, and rock and roll. We experience pleasure thanks to the particular neurotransmitter chemicals that are being released in the brain and to the particular regions of the brain where they are being released.

The function of each neurotransmitter depends entirely on the function of the structure in which it is located. Let us look at a few examples. Deep within your brain is a region called the basal ganglia. The neurons in the basal ganglia are responsible for producing normal well-controlled smooth movements. The level of the neurotransmitter dopamine in these nuclei is much higher than in most surrounding brain regions. Therefore, scientists have concluded that dopamine within the basal ganglia is critically involved in the control of movement. Furthermore, if we expose your brain to a drug that impairs the function of dopamine, then your ability to move will be impaired. Dopamine is obviously critical for movement. It would be incorrect, however, to assume that dopamine is only involved with the control of movement—it is not. You also can find dopamine in the retina of your eye and in your hypothalamus, structures that have nothing to do with movement. Dopamine also is released into small regions deep within the frontal lobes; its release produces a feeling of pleasure. Similarly, the neurotransmitter norepinephrine can be found in the hippocampus, a brain region that is critical for forming new memories. Thus, norepinephrine influences the formation of memories. However, norepinephrine also plays a role in other brain regions that have nothing to do with making memories. The take-away point is that there is no such thing as a specifically unique "dopamine function" or an exclusively distinct "norepinephrine function." The brain region within which the neurotransmitter is found defines its function, not the neurotransmitter itself. In fact, neurotransmitters exhibit such a complex array of actions in various brain regions that we can rarely make a single universal statement about their role in brain function.

Neurotransmitter chemicals are produced from the contents of your diet; thus, what you eat can, under certain conditions, influence how you think and feel. First, nutrients such as amino acids, sugar, and fats are absorbed from your food and transported across the blood–brain barrier into your brain; these nutrients are then absorbed into your neurons, where specialized enzymes convert them into neurotransmitters. The neurotransmitter molecules are then stored in very tiny spheres, which sit patiently waiting for the arrival of an action potential that instructs the neuron to release them. Once outside the neuron, the neurotransmitter wanders around looking for a way to communicate with the next neuron. The junction at which two neurons communicate is a synapse. The neurotransmitter molecule, now free to wander around within the synapse, will ultimately bump into and connect with a special protein, called a receptor. Receptors are like boats floating on the outer surface of the neuron on the other side of the synapse. Receptors offer comfortable docking ports for the neurotransmitter to insert itself. Once this docking of neurotransmitter and receptor has been achieved, the next stage in the communication process between neurons begins. At this point, lots of different things could occur; ions might move in or out of pores, enzymes might be activated, genes might be turned on or off, and many other possibilities. These secondary processes may have long-term consequences for the neuron's behavior and ultimately for your thoughts and actions.

Meanwhile, back in the synapse, after interacting with the receptor, the actions of the neurotransmitter must be terminated by means of its reabsorption back into the neuron that originally released it. This vacuuming-up process is called reuptake. Alternatively, the neurotransmitter also might be acted upon by local enzymes and converted into a chemical that can no longer interact with your brain. Once the neurotransmitter is inactivated, it is removed from the brain into the bloodstream. Such byproducts of the ordinary hustle and bustle of the brain can be monitored easily in many of our body fluids, and this information can be used to determine whether our brains are functioning normally. One thing that does not happen: Neurotransmitters produced in the brain do not leave the brain intact. They either are metabolized or their escape is blocked by the blood–brain barrier. This blockade is crucial because if these neurotransmitters escaped from the brain they would change, possibly with lethal or unpleasant consequences.

In summary, at the level of individual neurons communicating with other neurons, the process is both electrical, via the passage of an electrical disturbance called an action potential traveling down the axon, and chemical, via the release of a neurotransmitter onto the next neuron. Imagine that when your phone rings the electrical signal traveling via the telephone lines outside your house that brought the call into your telephone is similar to an action potential traveling down an axon. Someone has sent you a signal. Now imagine that you pick up the telephone receiver, you hold it to your ear, and the phone spits some chemicals into your ear. Your ear is the receptor for the chemical. This is how the brain works at the level of one neuron communicating with another neuron. Everywhere in your brain, one neuron is being electrically induced to spit chemicals at its neighbor. The chemicals released are obtained from your diet; this offers some insight into how important your diet is to normal brain function. This topic will be discussed in greater detail later. Now that you are familiar with the individual components of the brain, neurons and glia, and the chemicals that they use to communicate with each other, let us put it all together. Now would be a good time to examine closely the drawing of the brain included at the front of this book.

How is my brain organized?

During the past few decades, with the introduction of noninvasive techniques to examine brain function, neuroscientists have resurrected, rather inadvertently, a description of brain function that resembles a discredited idea from 1796. In that year a German physician, Franz Gall, developed an approach to understanding the brain by focusing on measurements of the human skull based on the concept that certain brain areas, or modules, have specific localized functions. The idea was intuitively attractive and became quite popular; in addition, the approach rather crudely described how the brain actually functioned. Today, over 200 years later, thanks to the invention of some highly sophisticated and expensive scanning machines, we have returned to the concept of a compartmentalized brain. Some parts of the brain clearly are dedicated to specific functions, such as vision, hearing, or touch; thus, the idea of modules is not entirely erroneous. A better analogy, however, is to view the brain as an orchestra that requires many unique instruments (brain regions) to work together simultaneously to produce a complex pattern of activity leading to the emergence of a beautiful piece of music (e.g., a perception of a blue bird flying across one's field of vision). Please keep this concept of an orchestra in mind as we now introduce the functions of various brain regions.

What happens in the front half of my brain?

Just behind each eye, in the front half of your brain, are the frontal lobes. Let us begin this discussion with the frontal lobes because they control such fascinating behaviors. Much of what is known about the function of these frontal lobes was learned by examining the behavior of people who suffered traumatic brain injury, stroke, cancer, or infection within that area of the brain. During the past few decades the use of noninvasive technologies, such as the magnetic resonance imaging (MRI) scanner, has provided the opportunity to monitor the activity of the frontal lobes in conscious humans performing specific tasks. This is what scientists have learned.

The frontal lobes allow you to make decisions, plan your actions, organize your thoughts about specific goals, and control your behavior around others. You inherit numerous personality traits from your parents, and these traits are controlled by selected regions within the frontal lobes. It is probably not too surprising to learn that variances in the size of some regions of your frontal lobes correlate with specific personality features. For example, introspection is correlated with the size of a region that lies on the top part of the frontal lobes called the prefrontal cortex. If you move your focus to a region just lateral to the part of the frontal lobe (just behind your eyebrows) responsible for introspection, you will find a region that becomes active when we perform complex behaviors such as paying attention or lying.

Lying is apparently a complex task that requires considerable attentional abilities and significant participation by this frontal brain region. The demand for such energetically costly cortical activation might explain why most of us are not very good liars; we may simply lack the piece of cortex necessary to pull it off successfully. Some individuals are born with cortical proclivities that allow them to be very good liars and also allow them to be successful in certain publicly supported professions.

If you now move your focus to the most lateral aspect of the frontal lobes, just inside of your temples and behind your eyes, you will find a region, the inferior frontal cortex, that is responsible for controlling low-risk behaviors. This region is in constant communication, and competition, with a region of the brain called the nucleus accumbens that induces you to participate in high-risk behaviors. These two regions might be seen as competing for control when you are trying to decide whether to have another piece of chocolate cake: the inferior frontal cortex is saying "No, you do not need another piece, it will just make you fat," while the nucleus accumbens is saying "Eat it! It will taste so good." Once your frontal lobes have made a decision, there is only one thing they can do—instruct some muscles to contract and move a part of your body. In spite of the complexity of our frontal lobes and the sophistication of their neural processes, our brain has a limited number of options—it can contract a muscle to move a limb or finger to pick up that second piece of cake. That is about all they can do in response to our wonderfully complex thoughts—move you from here to there, and back again.

You live in the frontal lobe of your brain. Yes, the *you* who is thinking about the meaning of the last sentence; the *you* who is feeling hot or cold right now, or hungry, or angry, or absolutely anything at all—that *you*. This particular region of cortex is part of a small circuit of brain systems that turn on as soon as you wake up in the morning. Think of that very familiar feeling you have immediately upon awakening that you know who you are and where you are in relation to familiar objects and other people. The frontal lobes are also one of the brain regions that become active when your mind is wandering or when you are simply recalling episodes from your past or contemplating actions in the future. Thanks to recent studies using sophisticated scanning machines, we now understand that the region of cortex on the medial (deep inside between both hemispheres) surface of the frontal lobes is responsible for allowing us to have a "Theory of Mind," or the ability to recognize that other people have thoughts and feelings that are independent of the *you* that lives just behind your eyes.

How does my brain produce speech?

Now that you have generated some thoughts, it is time to tell your friends what you are thinking. Speaking begins with the frontal lobes. Neurons in the frontal lobes control all of your muscles, including the muscles of the mouth necessary for speech. On the left side of your frontal lobe is a brain region, called Broca's area, which controls the actions of the muscles in your face and throat that allow you to produce sounds that have meaning to the people listening to you, that is, to produce speech. This small brain region was named after the French physician Paul Broca (1824–1880), who first identified a patient with damage to this part of the brain. Unfortunately, Broca's area lies in a region of the brain that is highly vulnerable to strokes. When Broca's area is damaged, speech becomes difficult, incoherent, or completely nonfluent. In contrast, people with damage to Broca's area can understand what is being said to them; this is because the part of the brain that is responsible for understanding language is in the back half of the brain. People with damage to Broca's area cannot generate normal speech. Speech is a complicated process that is not entirely understood. Neuroscientists have discovered that people who are deaf and use sign language to communicate also are impaired by damage to Broca's area. Overall, the job of the frontal lobes is to decide what you wish to say and then plan the actions of the muscles in the throat and mouth, with the help of neurons in Broca's area, to produce the sounds that can be interpreted by another person as something meaningful, that is, as speech. An important general rule for the brain is that the amount of cortex devoted to a particular function is related to the complexity of the function. Broca's area is, therefore, quite large in humans because speech is a complicated thing to produce. Obviously, because the size of the cortex is finite, there are limits to how much processing the cortex can simultaneously perform. The generation of speech is so demanding for our frontal lobes that talking to someone while driving draws mental resources away from driving and leads to a noticeable deterioration in driving performance. Similarly, focusing our attention on visual imagery reduces our ability to detect sounds.

What happens in the back half of my brain?

The remaining parts of the cortex are devoted to receiving, interpreting, and storing sensory information received from the eyes, ears, nose, tongue, skin, muscles, and joints. Let us begin with your eyes. How do you see? The occipital cortex lies at the back of your head. It is dedicated to processing information coming from the eyes. Both eyes send information to both halves, or hemispheres, of your brain. The occipital cortex "sees" lines; that is all, just lines. Amazingly, that is enough. After all, every object in your field of view is essentially lots of very short lines connected together to make letters, faces, and shapes. Within the occipital lobe, the world "looks" like a 1950s-era television show; that is, it is in black and white. At this level, the information on color that was provided by the retina is still being processed separately from the information about lines that make up shapes and faces. Currently, no one knows with certainty where in the brain the color of an object and its shape are merged to produce the typical visual image of the world around us.

The occipital lobe, along with some nearby regions, is responsible for some truly amazing visual gymnastics to make sure that your visual world behaves as you expect it to behave. For example, your eyes are constantly moving as you survey your environment. Take a moment and watch someone else's eyes as they dart back and forth: This is called saccadic eye movement. The other person is not even aware that his eyes are doing this; of course, your eyes were doing the same thing while you were looking at him. A saccadic eye movement shifts the image of the object you are looking at from one part of the retina to another. As you read this sentence, your eyes are bouncing around very rapidly. So, why doesn't your visual world appear to bounce around as quickly as your eyes are jumping? How does the brain manage to make the world appear so smooth and quiet? The solution is quite simple: The brain does not pay attention to any signals coming from the eyes when they are making these fast saccadic movements. During each saccadic jump, your brain stops processing visual images for just an instant and makes you believe that nothing happened. Your brain preserves your personal instant-by-instant visual representation by merging information obtained before and after each saccade. By doing this, the brain fills in your visual reality with a continuation of what you were seeing just before the visual image on your retina jumped. This sleight of hand by the brain gives us all the impression that the world is a smooth, flawless, and continuous flow of images. This deception is identical to what happens when you are watching a movie. You are aware that the movie is a sequence of photographic stills being flashed quickly up on to the screen in front of you, and yet you experience the movie as a smooth flow of images. Saccadic eye movements are occurring constantly; indeed, they occur so often that you are essentially not "seeing" anything at all for 60% to 70% of the time! Yes, two thirds of the time while you are awake and your eyes are wide open, you are effectively blind; you just never notice it. Your brain is incredibly good at fooling your mind into thinking that nothing happened and that the world is an uninterrupted, endless flow of images.

In contrast, you are aware that you are not seeing anything for a small fraction of a second when you blink. Blinks last longer than saccades; thus, the brain has time to take advantage of this short pause of incoming electrical signals and activate a unique pattern of brain regions called the "default network." These same brain regions also become engaged when you are by yourself, undisturbed and bored, and your mind begins to wander. It is hard to believe that during the time it takes you to blink, your brain attempts to go off-line and daydream, but it does. Sometimes your mind wanders for much longer periods of time.

Why does your mind wander? Why does your brain find it so easy to reconsider distracting thoughts from the day or to speculate on potential conversations you might have tomorrow? Why can't your brain just lie still and be quiet? Anyone who has ever tried meditation has discovered how incredibly difficult it is to quiet the mind—it takes lots of practice to be successful. Simply stated, your brain did not evolve in a world that rewarded you for being completely still and without thought. Any organism with a brain that completely disengages itself regularly would soon find itself being digested by a bigger organism not distracted by self-reflection. I could be wrong, but I doubt that *Tyrannosaurus rex* was a very pensive creature.

Why do I daydream?

Surprisingly, the answer has everything to do with why humans enjoy ingesting coffee and cocaine. Brains really like stimulation—thoughts or drugs, it really does not seem to matter. When you do not provide your brain with input from the external world, such as TV, music, or exploring social media sites, your brain actively disengages and starts to produce its own stimulation, namely, daydreaming. So when you are sitting through a boring lecture or listening for the hundredth time to your uncle tell that story about the big fish that got away, your mind has a tendency, sometimes an urgency, to go offline and entertain itself with other thoughts that it finds more interesting, such as "What will you have for dinner tonight?" or "When did your sister grow that mustache?" Neuroscientists now have identified the brain regions that selectively turn on, as well as those that turn off, when people are daydreaming; these include regions in the frontal and parietal lobes. These studies also have determined that daydreaming is important for normal brain function; daydreaming helps you to sort out important thoughts and discard nonsense and worries. Recently, scientists have estimated that while you are awake you spend approximately 60% to 70% of the time daydreaming! So, overall, most of the time that we are awake during the day we are either blind (due to saccadic eye movements) or completely offline (daydreaming). It is amazing that we ever get anything accomplished.

Why does the brain devote so much time to daydreaming?

Your brain evolved in a sensory rich world and rewards you for exposing it to ever more complex sensory experiences. Every time you experience something new, your brain releases a jolt of dopamine in the frontal lobe; dopamine is the major reward-related neurotransmitter in the brain. You can stimulate the release of dopamine in your brain artificially by ingesting coffee and cocaine or you can turn on the TV, listen to music, have sexual relations, or simply communicate with someone else important to you. The brain has evolved a system that rewards you for obtaining new information and having lots of thoughts because doing so might have survival value. The more you know, the more likely you are to survive and pass on your inquisitive genes to the next generation. Thus, you are burdened, or possibly blessed, with a brain that demands constant entertainment, via its own thoughts or exogenous chemicals, and that powerfully rewards you, by releasing dopamine, for providing it.

What is the function of the temporal lobe?

Moving your attention forward, the parts of your brain that lie directly underneath your ears are the left and right temporal lobes. As discussed in the first chapter, different parts of the temporal lobe are devoted to processing memories, the recognition of objects such as chairs and faces, and responding to these memories and objects with emotion. Damage to the bottom half of the temporal lobe can prevent someone from recognizing familiar faces and objects. For these unlucky people everything appears to be unfamiliar, even though they are certain the situation is familiar. This is called *jamais vu* from French meaning "never seen." The condition often involves a sense of eeriness; the patient feels as though she is seeing the situation for the first time even though she has been in the situation before.

The temporal lobe also is devoted to understanding the meaning of information coming from your ears. This region is called the auditory cortex and it is organized just like the keyboard on a piano, with the high notes at one end of the temporal lobe and low notes at the other. Damage to the auditory cortex removes a person's awareness of sound, but he can still respond reflexively to startling noises. This is because your response to unexpected or loud sounds is handled by a very primitive region of the brain that is not part of the normal auditory cortex. The region of temporal lobe that is dedicated to understanding language (either spoken or signed) is found only on the left side (for most people) of the brain and is called Wernicke's area. Damage to this area causes a condition called aphasia. Patients injured in this area are unable to understand language in its written, spoken, or signed form. The language of patients who experience injury or stroke in this area has a normal rhythm, but it does not make sense. Patients who ultimately recover their language abilities report that when they were impaired they found the speech of others and themselves to be completely unintelligible. These patients do retain the ability to sing and to utter profanities; these abilities are believed to be processed by regions of cortex in the opposite hemisphere (usually the right hemisphere). The left hemisphere is often called the dominant hemisphere due to the presence of language in this half of the brain. The nondominant right hemisphere, however, also pays attention to language. The same regions in the right hemisphere may play a role in the simultaneous processing and resolution of lesser meanings of ambiguous words. When speaking English, we all struggle with the fact that some words in English have multiple meanings; it is the job of the right hemisphere's temporal lobe to sort this out.

What is a seizure?

The circuitry and cellular architecture of the temporal lobe make it particularly vulnerable to generating patterns of neural activity that lead to seizures. A seizure occurs when neurons within a small brain region spontaneously fire off a series of electrical signals to neighboring neurons recruiting them into a self-sustaining pattern of recurrent and increasing activity that then spreads to other nearby brain regions. You might have witnessed the human version of this effect while sitting in a baseball or football stadium. A small group of people in the stands stood up and threw their hands skyward, thus inducing the people in a nearby section of the stadium to do the same. Slowly at first, then faster and involving more and more people, this wave of standing and waving fans crawls around the circular stadium until "the wave" becomes a self-sustaining rhythm of energy powered by the needs of people watching a ball game to do something to relieve the boredom of just sitting quietly. Brains behave in much the same way. Often, however, the original cause of the spontaneous activity is not completely understood; traumatic brain injury, fever, brain infections such as encephalitis and meningitis, stroke, brain tumors, and a variety of genetic syndromes may all contribute to the development of seizures. Neurons often behave as though they prefer to fire synchronously with each other rather than remain inactive. The problem is that the presence of seizure activity makes the normal functioning of the brain nearly impossible, just as participating in "the wave" during a ball game makes paying attention to the action on the field of play difficult. Once a seizure begins, it follows the natural behaviors of neurons; that is, it spreads to neighboring brain regions in just the same way that all neuronal information flows in the brain. As the seizure recruits more and more neurons, it spreads across the cortex as a wave of electrical disturbance crawling from one region to the next until the entire brain becomes engaged.

Seizures always should be treated aggressively and never be allowed to occur repeatedly. Why? Because the increased neural activity in the brain that characterizes a seizure is very similar to the conditions that underlie the process of learning; ultimately, as seizures continue to occur, the brain will learn to produce more and more seizures. Usually, the neural pathways that are contributing to the seizures involve the neurotransmitter glutamate; the increased release of this amino acid in the brain leads to neuronal cell death. Therefore, repeat seizures must be avoided so that seizure incidence and cell death are reduced. The unique distributed locations of specific brain functions become evident by observing someone having a seizure. For example, the onset of a seizure originating within the temporal lobe is often associated with the function associated with that same temporal lobe region. Patients may report amnesia or the recall of a specific memory, an abnormal taste or smell, an extremely sad or happy unexplained emotion, or an auditory or visual hallucination. These are all temporal lobe processes being initiated by the uncontrolled activation of neurons that characterizes a seizure episode.

Epilepsy is a common brain condition that has likely plagued humans for many millennia. As mentioned earlier, the temporal lobe is particularly vulnerable to generating seizures. A recent investigation discovered that the tendency to display extravagant religious behaviors correlated significantly with pathology within the right temporal lobe of patients with untreatable epilepsy. In fact, the medical literature is replete with reports of epilepsy patients who demonstrated elaborate religious delusions. Seizures can induce vivid hallucinations that often take on other-worldly or religious features. When auditory hallucinations occur, they usually involve a single voice at a time, speaking in the native language of the patient.

What is a hallucination?

Studies of hallucinations have demonstrated that your experience of the world is not solely determined by the direct sensory inputs into your brain. When you perceive something but there is nothing there to perceive, you are hallucinating. When your brain completely misinterprets incoming sensory information, you are hallucinating. When your brain generates its own sensory inputs without any assistance from the normal sensory systems, you are hallucinating. Any sensory modality—visual, auditory, tactile, and gustatory or even a mixture of these, called cross-modal—can experience a severe misinterpretation of the neural signals bouncing around in the back half of your brain. This is why seizures can induce hallucinations; they are characterized by random neural activity that your brain attempts to interpret as something meaningful. No one is certain how hallucinations are initiated in the brain. People with schizophrenia or dementia frequently report auditory or visual hallucinations; the severity of the illness correlates with the presence of visual hallucinations in these patients. Some theories suggest that these hallucinations occur because schizophrenic patients lack the ability to distinguish self-generated from externally generated thoughts and sensory experiences. If you cannot tell the difference between what your brain is generating versus what is really happening, you also may lose the ability to predict the consequences of what you are about to perceive. This odd situation has an unusual consequence that is quite fascinating: if you cannot predict the consequences of your movements, you should, unlike everyone else, be able to tickle yourself. Astonishingly, schizophrenics can tickle themselves. This peculiar trait offers insight into which brain regions are not functioning normally in the brains of schizophrenics.

In order to gain some insight into what it means to hallucinate, we might begin by examining what pharmacological hallucinogens do inside our brains. One of the best studied hallucinogenic drugs, and the topic of numerous bad movies from the 1960s, is lysergic acid diethylamide (LSD). Once ingested, LSD attaches itself to a variety of serotonin receptor proteins all over the brain. Serotonin is a neurotransmitter released by neurons that project their axons to every part of your brain. If you were able to insert a recording device into a serotonin neuron, you would discover that it has a regular, slow spontaneous level of activity that varies little while you are awake. When you fall asleep, the activity of these neurons slows. When you start to dream—or if, as we will see shortly, you ingest a hallucinogen—these neurons cease their activity completely. The effects of LSD on serotonin neurons may be the initial trigger that sets in motion a cascade of complex processes throughout the brain that is experienced as a hallucination. In truth, no one currently understands how LSD or any of the hallucinogens actually work, or just how serotonin factors into their hallucinatory effects. Confounding this uncertainty is the fact that some powerful hallucinogens, such as Salvinorin A, have no effect on serotonin function at all. Some stimulants, such as amphetamine and cocaine, produce sensory hallucinations via the activation of dopamine receptors; others, such as PCP or ketamine, act directly on glutamate receptors. Hallucinations sometimes occur in people who live normal healthy lives and are never aware that their peculiar sensory experience is not shared by everyone. This is called synesthesia.

What is synesthesia?

Imagine yourself as a newborn baby lying in a crib. Your brain's serotonin neurons at this age are not fully functional and your serotonin receptors have not yet converted to an adult profile. Your sensory systems—eyes, ears, fingers, toes, and nose—are all working quite well, but due to the immature state of your serotonergic system, your brain does not correctly process all of the incoming sensory information. Thus, visual information is blended with auditory signals, smells are confused for colors, and touch produces a sound. Why does this happen? One theory is that normal serotonin function is required for the brain to accurately process sensory information in the cortex. If this neurotransmitter system is not working correctly, then sensory experiences become confounded with each other. This experience is called synesthesia. Some scientists theorize that when we are only a few days or months old synesthesia is a normal experience for the brain due to the immature state of brain chemistry and anatomy. Twenty years later, with a fully functioning mature serotonin system, consuming LSD induces a temporary synesthesia experience similar to the one you had in your crib as an infant. Why? The inhibited function of your serotonergic system that is induced by LSD, or possibly any hallucinogen, may reproduce the condition of synesthesia that was "normal" when you were a newborn. As a newborn, you found this condition to be frightening and you cried. After all, who wouldn't? However, as an adult who has just ingested some LSD, you might, in the right setting, come to believe that the condition is a transcendently mystical experience. In fact, it is not mystical; it is a drug-induced replication of the conditions that originally existed in your infant brain prior to the maturation of a small group of neurons that release serotonin. Fortunately, due to your "infant amnesia," you remember nothing of this bizarre experience. Keep this in mind next time you find yourself hovering over the crib of a crying frightened baby—you may be inducing some terrifying hallucinations for the child.

Some people never grow out of this infant phase of constant hallucination. These people are synesthetes. Imaging studies have found that these individuals have abnormalities that are consistent with altered patterns of connectivity within various temporal lobe brain regions. Synesthesia has a strong genetic component; it runs in families and males get it about as often as females. However, not everyone who gets the variant gene shows the symptoms. The condition is currently considered a harmless alternative form of perception.

What happens in the parietal lobe?

Next, move your attention to the top of the back half of the brain, which is called the parietal lobe. The front part of this lobe is responsible for processing the sensations of touch and taste. The back half of the parietal lobe is one of the most recently evolved brain regions and has a very complex task: it is responsible for integrating the sensory information (primarily visual information for humans) coming into the brain from all over the body into a single "world view" that is unique for each person. The parietal lobe also receives information from the frontal lobe. For example, once your frontal lobes have decided to move your right arm, a copy of this decision is sent back to the parietal lobe; the parietal lobe uses this information to predict the sensory consequences of the impending movement. The parietal lobe always wants to know what you are going to do before you do it so that it can anticipate the arrival of the subsequent sensory experience. This is why you cannot tickle yourself; it is impossible to sneak up on a region of your body because your parietal lobe always knows what your hands are doing even before they do it.

Imaging studies have revealed that the parietal lobe becomes active when we are envisioning the future, making moral decisions, or recalling autobiographical memories. The complexity and abstract nature of these tasks may explain why the parietal lobe evolved so recently in vertebrates. It is probably not surprising that the parietal lobe is a vital component of the default network involved with daydreaming. Of all the lobes in your brain the parietal lobe is probably the least understood. Recent studies indicate that the parietal lobe shows significant pathology during the early phases of Alzheimer's disease, which likely contributes to some of the initial diagnostic symptoms such as confusion, delusion, disorientation, and difficulty thinking, understanding, and concentrating.

Where is my cingulate gyrus and why should I care?

Imagine the brain as a melon you are holding with both hands. Now slice the melon down the middle so that you are holding two identical halves in each hand. Each half is analogous to one brain hemisphere. Now look at the inner flat surface of one of those halves; you would be able to see another important brain region, the cingulate gyrus, running horizontally from the front to the back of the brain. This gyrus performs some quite interesting tasks. Studies using noninvasive scanning machines have discovered that the cingulate gyrus becomes active when we experience either social or physical pain, thus confirming what all of us have always known—words can cut like swords and produce true misery. This gyrus is also active while making decisions about your next behavior. The cingulate gyrus may help you decide whether your next behavior will be rewarded or punished. This brain region is part of a large group of regions, collectively called the limbic system, which was discussed in Chapter 2.

The volume of the cingulate gyrus is greatly reduced in the brains of patients with bipolar disorder or major depressive disorder. It is currently unknown whether the shrinkage of this critical brain region precedes the onset of the symptoms of these illnesses or is a consequence of experiencing long-term depression. In addition to its potential role in depression, recent imaging studies have found that the anterior region of the cingulate gyrus is smaller in schizophrenic patients and that this change correlated with a lower level of social functioning and a higher degree of psychopathology. Whether this correlation is the source of these symptoms remains to be determined by future studies.

A colleague once had a patient with a benign tumor growing between her two hemispheres; as the tumor grew in size, it began to push against the cingulate gyrus in both hemispheres of her brain. The tumor was discovered during an examination following a car accident when the patient complained that she developed some unusual symptoms. The most problematic symptom was that she lost the ability to control her sexual desires. This was most distressing to the patient because she was a cloistered nun. Fortunately, when the tumor was ultimately removed, the woman was once again able to control her urges and return to her chosen vocation.

What is "the little brain" and what does it do?

Long ago, anatomists saw what appeared to be an additional companion brain hanging off the back of the larger hemispheres and decided to call it the cerebellum, or "little brain." It is a tennis-ball-sized structure, only about a tenth of the size of the brain that, surprisingly, contains almost 50% of all of the neurons in your head. It has a highly convoluted cortex with only three layers; the interior of the cerebellum contains lots of myelinated axons going to and from the brain and spinal cord. What does the cerebellum do? Once again the answer to this question comes from studies using noninvasive scanning machines and indwelling electrodes. The cerebellum plays a role in the control of certain types of memory and mood. We also know that the neurons within the cerebellum become active just prior to and during the contraction of the muscles of the body. When the cerebellum is damaged due to injury, stroke, or tumors, the most common symptom is difficulty with movement and posture. People with cerebellar damage can move, but their movements are not smooth or well controlled. The cerebellum receives sensory information from your muscles and joints to inform you about the location of your body parts; this allows you to move correctly without paying attention to every movement. Patients with cerebellar damage usually find that they need to walk with their feet placed widely apart and, because they cannot tell where their limbs are located, they need to watch what their limbs are doing at all times.

The ability of the cerebellum to perform complex well-learned movements smoothly also can be impaired by alcohol or marijuana intoxication; this is because the cerebellum contains specific receptor proteins that respond to both of these drugs. If you are ever stopped by a police officer on suspicion of driving drunk, you may be asked to touch your nose with your outstretched finger. Ordinarily this is quite an easy task to perform; however, it is not easy to perform when the cerebellum is bathed in alcohol. Alcohol distorts the pattern of neuronal activity, preventing you from moving your arm accurately to touch your nose. Alcohol also distorts the ability of your cerebellum to control the smooth coordination of the muscles of the eyes. When the police officer instructs you to follow his finger with your eyes, he is testing to determine whether alcohol has impaired the ability of your cerebellum to control the muscles of your eye. If your eyes begin to move involuntarily from side to side in a rapid swinging motion rather than staying fixed on the officer's finger, you are displaying what is called nystagmus. Nystagmus can be induced by alcohol intoxication. Obviously, the little brain in the back is just as important as the big brain up front when it comes to our survival and the success of our species.

How does my right brain talk to my left brain?

The two hemispheres of your brain talk to each other via the axonal highway made of axon fibers called the corpus callosum. Surprisingly, the size of the corpus callosum gets proportionally smaller as the cortex increases in size. Why? The reason is that decreasing the number of connections between hemispheres allows each half of your brain to spend less effort integrating its activity with the other half; this adaptation makes it possible for each half of your brain to specialize in specific abilities. For example, the part of the brain that controls the production of speech is solely in the left (usually) hemisphere. The advantage is that small regions of cortex in just one hemisphere can be dedicated to a particular function, thus allowing the same region in the other hemisphere to handle a different ability. This allows your brain to control a much broader range of abilities. The delegation of tasks to different hemispheres has many advantages.

There is also one very big disadvantage; small injuries in one hemisphere, due to a stroke or cancerous growth, can produce very big deficits in function. For example, a small stroke in just the left hemisphere might destroy your ability to talk. The distribution of specific abilities to a single hemisphere has made our species very broadly skilled and improved our chances of survival as an individual and species; the cost is that we are very vulnerable to small injuries.

Final thought

At the beginning of this book, I stated that my purpose was to provide the most accurate and up-to-date information possible about the brain. If I have achieved my goal, then you have gained some insights into how your brain is organized, how it evolved, and how it generates emotions, hallucinations, and memories. I hope that you also have learned that there is a degree of predictability in how your brain responds to drugs and the food you eat and that what you eat does influence how you think and how fast you age. I encourage you to continue learning more about your fascinating brain from the suggestions for further reading that are provided here. I promise that the more you learn about your brain, the more determined you will become to understand how your mind emerges from the three pounds of tissue floating in your head.

GLOSSARY

Acetylcholine A neurotransmitter in the brain and body that is formed by the combination of acetic acid and choline.

Acetylcholinesterase An enzyme present in nervous tissue and at the neuromuscular junction that catalyzes hydrolysis of acetylcholine to choline and acetic acid.

Action potential A brief change in the electrical charge across the membrane of a nerve that travels away from the cell body, along the axon, until it reaches the axon terminal, where it induces the release of a neurotransmitter.

Alzheimer's disease An irreversible and progressive brain disorder that slowly impairs memory and thinking skills, and eventually the ability to perform simple tasks.

Amnesia An inability to remember past events or impairment in the ability to form new long-term memories.

Amygdala A structure located in the ventral region of the temporal lobe and considered to be a part of the limbic system. It is involved in the initiation of emotions, principally the production of a fear response.

Analgesic A drug that dulls the sensation of pain. It differs from an anesthetic agent in that it relieves pain without loss of consciousness.

Anticholinergic A drug that blocks, inhibits, or antagonizes the actions of acetylcholine or other cholinergic receptor agonists. Because the predominant effects of the parasympathetic nervous system are mediated by acetylcholine, the term "anticholinergic effects" often is used to imply an inhibitory action in the parasympathetic nervous system.

Anticonvulsant A drug that reduces the incidence of convulsions or seizures.

Anxiolytic A drug that reduces anxiety.

Astrocytes One of the three major classes of glial cells found in the central nervous system; important in providing nutrients to the neurons and removing metabolic and neurotransmitter byproducts.

Autonomic nervous system Subdivided principally into the sympathetic and parasympathetic efferent systems. It controls the function of visceral organs and allows the expression of the physical components of emotion.

Axon. The straight, relatively unbranched process of a nerve cell that carries the action potential to the terminal for release of a neurotransmitter.

B.C.E. Before the common era.

Bilateral On both sides of the brain.

Blood–brain barrier Functional barrier produced by glia wrapped around blood vessels preventing access for many blood-borne molecules to the brain. The barrier consists mostly of the fatty membranes of the astrocytes and the existence of tight junctions between the vascular endothelial cells.

Broca's area An area in the left frontal lobe specialized for the production of speech.

Central nervous system The brain and spinal cord.

Cerebellum Means "little brain" in Latin. It is involved in learning, the coordination and production of speech, the organization of muscle movement, coordination of fine motor movement, and balance; it is the center of a feedback loop involving motor and sensory information.

Cerebral cortex The thin outer layer of the cerebral hemisphere, which contains neurons that are organized in six horizontal layers as vertical columns. It appears as ridges (gyri) and narrow folds (sulci) in order to maximize the number of neurons and number of columns within the confined space of the skull. It is responsible for all forms of conscious experience, including perception, emotion, thought, and planning. Cortex means "bark" in Greek; the bark of the cork tree looks a lot like the cerebral cortex.

Cerebral hemispheres The two halves of the brain. The left hemisphere is specialized for initiating speech, language, writing, and calculation. The right hemisphere is specialized for initiating spatial abilities, face recognition in vision, and some aspects of music perception and production.

Corpus callosum A large bundle of axon fibers that allows the two hemispheres of the brain to talk to each other. The two hemispheres use this axonal highway to coordinate the sensory processing and planning that takes place in both hemispheres.

Dementia Loss of higher intellectual function. This condition may be progressive due to some underlying disease process or may be drug induced.

Dendrite The parts of the neuron that receive information from other neurons. These structures contain receptors and are able to form synapses with incoming neurons.

Dura mater The thick external covering of the brain and spinal cord; one of the three components of the meninges, the other two being the pia mater and arachnoid.

Endocannabinoids The endogenous lipid-soluble chemicals anandamide and 2-AG that bind to the brain's marijuana receptors. Endocannabinoids are unlike most of the brain's transmitters because they are not stored in synaptic vesicles but are synthesized by neurons upon demand.

Endorphins Protein neurotransmitters that mimic the action of morphine.

Frontal lobe One of the four lobes of the brain; it includes all of the cortex that lies in the front half of the brain.

GABA Gamma-amino butyric acid is a neurotransmitter with major inhibitory function in the brain and body. Its primary function is to turn off the activity of other neurons.

Gastrointestinal system Pertaining to the stomach and intestines.

Glutamate A neurotransmitter with major excitatory function in the brain and body. Its primary function is to increase the activity of other neurons.

Gray matter The parts of the nervous system that contain neuronal cell bodies and very little myelin (known as white matter). In fresh dissection, these regions appear grayish.

Gyrus A ridge or fold between two clefts (sulcus) on the cerebral cortex. The gyri and sulci create the folded appearance of the brain in humans and other mammals.

Hallucination An altered sensory experience in a conscious and awake state that can be initiated by the brain or due to the presence of an external stimulus, such as a drug. Hallucinations are most often visual and/or auditory, but also can involve the sense of taste, touch, and pain.

Hippocampus A structure found in the temporal lobe that is part of the limbic system and is important for the formation of new memories.

Ipsilateral On the same side of the body.

Hypothalamus A brain region located at the bottom center of the brain. It is responsible for controlling feeding behaviors, body temperature, thirst, and hormone release from the pituitary gland.

Limbic system A circuit of brain structures that plays a role in the control and production of emotional behavior.

Lipid solubility The property of a chemical to dissolve in fat. Due to the fatty components of the blood–brain barrier, lipid-soluble chemicals will enter the brain more easily than chemicals that are not lipid soluble.

Long-term potentiation (LTP) An increase in size of a synaptic electrical potential lasting one hour or more. LTP is considered an artificially produced electrophysiological representation of the actual neural processes that underlie a memory.

Mammal An animal the embryos of which develop in a uterus and the young of which begin to suckle at birth (technically, a member of the class Mammalia).

Microglial A type of glia that plays a critical role in the immune responses of the brain.

Mitochondria A small membrane-bound organelle found in large numbers in most cells of the body, in which the biochemical processes of respiration and energy production occur.

Myelin A sheath of fatty material that surrounds most axons. It acts as an insulator to enhance electrical conduction of action potentials.

Myelination Process by which glial cells wrap axons to form myelin that increase axonal conduction velocity.

Narcolepsy A rare disabling hypersomnia disorder that may include cataplexy, sleep paralysis, hypnagogic hallucinations, and sleep-onset rapid eye movement (REM) periods, but also disrupted nighttime sleep by nocturnal awakenings, and REM sleep behavior disorder (RBD).

Neuron The basic functional unit of the nervous system, also called a nerve cell. It is specialized for the transmission of information and characterized by long fibrous projections called axons, as well as shorter, branch-like projections called dendrites.

Neurotransmitter A chemical substance produced within neurons from components of the diet and released by neurons to diffuse into the extracellular environment and bind to specific postsynaptic protein receptors on a neighboring neuron.

Non-rapid eye movement (non-REM) sleep Collectively, those phases of sleep characterized by the absence of rapid eye movements.

Nystagmus An involuntary, usually lateral, back-and-forth rhythmic eye movement.

Occipital lobe One of the four lobes of the brain. It is responsible for vision and visual object and face recognition. This lobe is located at the most posterior part of the brain.

Orexin A protein neurotransmitter that controls arousal and feeding behaviors.

Parasympathetic nervous system Part of the autonomic nervous system. This system is involved with maintenance of bodily activities and conservation of energy.

Parietal lobe One of the four lobes of the brain. It is responsible for processing higher sensory and language functions. This lobe is located on the top of the brain.

Parkinson's disease A degenerative disease that results in a tremor at rest, usually involving the hands and feet at first, along with a general slowness in movement.

Pia mater The innermost layer of the membranes surrounding and protecting the brain that closely follows the bumps and wrinkles of the brain's surface.

Placebo A medicine or preparation with no pharmacologic activity, which is effective only by virtue of the power of suggestion associated with its administration.

Poikilotherm An organism whose internal temperature varies in relation to the temperature of the environment. It is the opposite of a homoeotherm, an organism that maintains thermal homeostasis.

Prefrontal cortex The very anterior-most part of the brain, which controls planning and thought.

Presynaptic The site from which the neurotransmitter is released, usually at the end of an axon.

Pseudoscience A claim, belief, or practice that is incorrectly presented as scientific but does not adhere to a valid scientific method, cannot be reliably tested, or otherwise lacks scientific status.

REM atonia A sleep disorder characterized by the loss of normal voluntary muscle tone during REM dream sleep.

REM behavior disorder Dream-enacting behaviors and movements during REM sleep. RBD is commonly associated with Parkinsonism and narcolepsy.

Receptor A protein floating on the surface of a neuron. Usually these proteins are the point of interaction between two neurons or between a drug and a neuron.

Reuptake When some neurotransmitters are released into the synaptic cleft, a fraction of the released molecules is recovered by transport back into the axon terminal, where it may be reused or degraded. Some drugs (called selective reuptake inhibitors) can interfere with this process.

Serotonin A neurotransmitter derived from the dietary amino acid tryptophan.

Sulcus (plural = sulci) The valleys or spaces between the folds or gyri of the brain.

Sympathetic nervous system The sympathetic nervous system is part of the autonomic nervous system. In general, this is a system that is involved with activation of bodily activities and mobilization of energy-consuming activities (increase in heart rate, respiration). This system has been characterized as the "flight-or-fight" system.

Synapse The place where one neuron connects to another neuron. The synapse includes the nerve terminal of the first neuron, the spot on the second neuron with receptors, and the space between them.

Synaptic vesicles Membrane spheres that contain neurotransmitter molecules that are stored near the presynaptic membrane at the synapse.

Synesthesia A phenomenon in which stimulation of one sensory pathway leads to automatic, involuntary experiences in a second sensory pathway. The experience varies in intensity and people vary in awareness of their synesthetic perceptions.

Temporal lobe One of the four lobes of the brain. It is responsible for processing hearing, olfaction, object recognition, language, speech, learning, and memory. It is located on lateral sides of each brain hemisphere near the ears.

Wernicke's area Region of cortex in the left temporal lobe that helps mediate language comprehension.

White matter Those parts of the brain and nervous system that primarily contain axons wrapped with myelin.

FURTHER READING

Chapter 1

Bliss TV, Collingridge GL (1993) A synaptic model of memory: long-term potentiation in the hippocampus. Nature, Vol 361, pp. 31–39.

De Leon J, Diaz FJ (2005) A meta-analysis of worldwide studies demonstrates an association between schizophrenia and tobacco smoking behaviors. Schizophrenia Research, Vol 76, p. 135.

Eichenbaum H (2008) Learning and Memory. New York: W. W. Norton & Company.

Giocomo LM, Hasselmo ME (2007) Neuromodulation by glutamate and acetylcholine can change circuit dynamics by regulating the relative influence of afferent input and excitatory feedback. Molecular Neurobiology, Vol 36, p. 184.

Gluck MA, Mercado E (2013) Learning and Memory: From Brain to Behavior, 2nd ed. Duffield, UK: Worth Publishers.

Hamann S (2005) Sex differences in the responses of the human amygdala. The Neuroscientist, Vol 11, p. 288.

Pfeiffer BE, Fostser DJ (2015) Autoassociative dynamics in the generation of sequences of hippocampal place cells. Science, Vol 349, p. 180.

Wenk GL (2003) Functional neuroanatomy of learning and memory. In: DS Charney, EJ Nestler, & BS Bunney (Eds.), Neurobiology of Mental Illness, 2nd ed., pp. 807–812. New York: Oxford University Press.

Wenk GL (2006) Neuropathologic changes in Alzheimer's disease: Potential targets for treatment. Journal of Clinical Psychiatry, Vol 67, p. 3.

Wenk GL (2014) Your Brain on Food: How Chemicals Control Your Thoughts and Feelings, 2nd ed. Oxford: Oxford University Press.

Chapter 2

Dravets WC, Price JL, Bardgett ME, Reich T, Todd RD, Raichle ME (2002) Glucose metabolism in the amygdala in depression: Relationship to diagnostic subtype and plasma cortisol levels. Pharmacology Biochemistry and Behavior, Vol 71, p. 431.

Falkai P, Rossner MJ, Schulze TG, Hasan A, Brzozka MM, Malchow B, Honer, et al.,(2015) Kraepelin revisited: schizophrenia from degeneration to failed regeneration. Molecular Psychiatry, Vol 20, p. 671.

Gazzaniga MS (2015) Tales from Both Sides of the Brain: A Life in Neuroscience. New York: Harper Collins.

Jamison KR (1996) An Unquiet Mind: A Memoir of Moods and Madness. New York: Vintagem.

Kramer P, Bressan P (2015) Humans as superorganisms: how microbes, viruses, imprinted genes, and other selfish entities shape our behavior. Perspectives on Psychological Science, Vol 10, p. 464.

Ladoux J (1998) The Emotional Brain: The Mysterious Underpinnings of Emotional Life. New York: Simon & Schuster.

Loftus EF (1979) The malleability of human memory. American Scientist, Vol 67, p. 312.

Lohoff FW (2010) Overview of the genetics of major depressive disorder. Current Psychiatry Report, Vol 12, p. 539.

Molgat CV, Pattan SB (2005) Comorbidity of major depression and migraine: A Canadian population-based study. Canadian Journal of Psychiatry, Vol 50, p. 832.

Muller AJ, Shine JM, Halliday GM, Lewis SJ (2014) Visual hallucinations in Parkinson's disease: theoretical models. Movement Disorders, Vol 29, p. 1591.

Patrick RP, Ames BN (2015) Vitamin D and the omega-3 fatty acids control serotonin synthesis and action, part 2: relevance for ADHD, bipolar disorder, schizophrenia, and impulsive behavior. FASEB Journal, Vol 29, p. 2207.

Perry E, Lee ML, Martin-Ruiz CM, Court JA, Bauman ML, Perry RH, Wenk GL (2001) Cholinergic activities in autism: abnormalities in the cerebral cortex and basal forebrain. American Journal of Psychiatry, Vol 158, p. 1058.

Posener JA, Wang L, Price JL, Gado MH, Province MA, Miller MI, Babb CM, et al (2003) High-dimensional mapping of the hippocampus in depression. American Journal of Psychiatry, Vol 160, p. 83.

Raison CL, Miller AH (2013) The evolutionary significance of depression in pathogen host defense. Molecular Psychiatry, Vol 18, p. 15.

Slow EJ, Postuma RB, Lang AE (2014) Implications of nocturnal symptoms towards the early diagnosis of Parkinson's disease. Journal of Neural Transmission, Vol 121, p. S49.

Standage T (2006) A History of the World in 6 Glasses. New York: Walker.

Tohyama M, Miyata S, Hattori T, Shimizu S, Matsuzaki S (2015) Molecular basis of major psychiatric diseases such as schizophrenia and depression. Anatomical Science International, Vol 90, p. 137.

Young JW, Dulcis D (2015) Investigating the mechanisms underlying switching between states in bipolar disorder. European Journal of Pharmacology, Vol 759, p. 151.

Zigmond MJ, Coyle JT, Rowland LP (2014) Neurobiology of Brain Disorders: Biological Basis of Neurological and Psychiatric Disorders. Salt Lake City, UT: Academic Press.

Chapter 3

Bendlin BB (2011) Effects of aging and calorie restriction on white matter in rhesus macaques. Neurobiology of Aging, Vol 32, p. 2310.

Benton D (2010) The influence of dietary status on the cognitive performance of children. Molecular Nutrition Food Research, Vol 54, p. 457.

Courtright DT (2001) Forces of Habit: Drugs and the Making of the Modern World. Cambridge, MA: Harvard University Press.

Fontana L, Partridge L (2015) Promoting health and longevity through diet: from model organisms to humans. Cell, Vol 161, p. 106.

Lane N (2005) Power, Sex, Suicide: Mitochondria and the Meaning of Life. Oxford: Oxford University Press.

Lane N (2016) The Vital Question: Energy, Evolution, and the Origins of Complex Life. New York: W.W. Norton & Co.

Linden D (2007) The Accidental Mind: How Brain Evolution Has Given Us Love, Memory, Dreams, and God. Cambridge, MA: Harvard University Press.

Marchalant Y, Cerbai F, Brothers HM, Wenk GL (2008) Neuroinflammation in young and aged rats: influence of endocannabinoids and caffeine. Journal of Neuroimmunology, Vol 197, p. 168.

Marchalant Y, Brothers HM, Wenk GL (2009) Cannabinoid agonist WIN-55,212-2 partially restores neurogenesis in the aged rat brain. Molecular Psychiatry, Vol 14, p. 1068.

Murakami K, Sasaki S (2010) Dietary intake and depressive symptoms: A systematic review of observational studies. Molecular Nutrition Food Research, Vol 54, p. 471.

Peneau S, Galan P, Jeandel C, Ferry M, Andreeva V, Hercberg S, Kesse-Guyot E (2011) Fruit and vegetable intake and cognitive function in the SU.VI.MAX 2 prospective study. American Journal of Clinical Nutrition, Vol 94, p. 1295.

Sánchez-Villegas A, Balbete C, Martinez-Gonzalez MA, Martinez JA, Razquin C, Salas-Salvado J, Estruch R, Buil-Cosiales P, et al (2011) The effect of the Mediterranean diet on plasma brain-derived neurotrophic factor (BDNF) levels: The PREDIMED-NAVARRA randomized trial. Nutritional Neuroscience, Vol 14, p. 195.

Stice E, Yokum S, Burger KS, Epstein LH, Small DM (2011) Youth at risk for obesity show greater activation of striatal and somatosensory regions to food. Journal of Neuroscience, Vol 31, p. 4360.

Wenk GL (2014) Your Brain on Food: How Chemicals Control Your Thoughts and Feelings, 2nd ed. Oxford: Oxford University Press.

Wenk GL (1991) Dietary factors that influence the neural substrates of memory. In: RL Isaacson, K Jensen (Eds.), The Vulnerable Brain and Environmental Risks. Vol 1: Malnutrition and Hazard Assessment, p. 67. New York: Plenum.

Wenk GL, McGann-Gramling K, Hauss-Wegrzyniak B, Ronchetti D, Maucci R, Rosi S, Gasparini L, Ongini E (2004) Attenuation of chronic neuroinflammation by a nitric oxide-releasing derivative of the antioxidant ferulic acid. Journal of Neurochemistry, Vol 89, p. 484.

Chapter 4

Cipriani G, Lucietti C, Danti S, Nuti A (2015) Sleep disturbances and dementia. Psychogeriatrics, Vol 15, p. 65.

Dauvilliers Y, Silber MH, Ferman TJ, Lin SC, Benarroch EE, Schmeichel AM, Ahlskog JE, et al (2013) Rapid eye movement sleep behavior disorder and rapid eye movement sleep without atonia in narcolepsy. Sleep Medicine, Vol 14, p. 775.

Frank MG (2012) Sleep and Brain Activity. Salt Lake City, UT: Academic Press.

Hasler BP, Bootzin RR, Cousins JC, Fridel K, Wenk GL (2008) Circadian phase in sleep-disturbed adolescents with a history of substance abuse: a pilot study. Behavioral Sleep Medicine, Vol 6, p. 55.

Kang JE, Lim MM, Bateman RJ, Lee JJ, Smyth LP, Cirrito JR (2009) Amyloid-beta dynamics are regulated by orexin and the sleep-wake cycle. Science, Vol 326, p. 1005.

Kredlow MA, Capozzoli MC, Hearon BA, Calkins AW, Otto MW (2015) The effects of physical activity on sleep: a meta-analytic review. Journal of Behavioral Medicine, Vol 38, p. 427.

Lockley SW, Foster RG (2012) Sleep: A Very Short Introduction. Oxford: Oxford University Press.

Luppi PH, O Clément, SV Garcia, F Brischoux, P Fort (2013) New aspects in the pathophysiology of rapid eye movement sleep behavior disorder: the potential role of glutamate, gamma-aminobutyric acid, and glycine. Sleep Medicine, Vol 14, p. 714.

Nixon JP, Mavanji V, Butterick TA, Billington CJ, Kotz CM, Teske JA (2015) Sleep disorders, obesity and aging: the role of orexin. Ageing Research Reviews, Vol 20, p. 63.

Peever J (2011) Control of motoneuron function and muscle tone during REM sleep, REM sleep behavior disorder and cataplexy/narcolepsy. Archives of Italian Biology, Vol 149, p. 454.

Stickgold R, Walker MP (2009) The Neuroscience of Sleep. Salt Lake City, UT: Academic Press.

Underwood E (2015) The final countdown. Science, Vol 350, p. 1188.

Chapter 5

Bardou I, DiPatrizio N, Brothers, HM, Kaercher RM, Hopp SC, Wenk GL, Marchalant Y (2012) Pharmacological manipulation of cannabinoid neurotransmission reduces neuroinflammation associated with normal aging. Health, Vol 4, p. 679.

Bass J, Takahashi JS (2010) Circadian integration of metabolism and energetics. Science, Vol 330, p. 1349.

Bausell RB (2009) Snake Oil Science: The Truth About Complementary and Alternative Medicine. Oxford: Oxford University Press.

Bendlin BB (2010) NSAIDs may protect against age-related brain atrophy. Frontiers in Aging Neuroscience, Vol 3, p. 2.

Boesen EH, Johansen C (2008) Impact of psychotherapy on cancer survival: time to move on? Current Opinion in Oncology, Vol 20, p. 372.

Brown WA (2012) The Placebo Effect in Clinical Practice. Oxford: Oxford University Press.

Claasen DO, JOsephs KA, Ahlskog JE, Silber MH, Tippmann-Peikert M, Boeve BF (2010) REM sleep behavior disorder preceding other aspects of synucleinopathies by up to half a century. Neurology, Vol 75, p. 494.

Duarte JMN, Schuck PF, Wenk, GL, Ferreira GC (2014) Metabolic disturbances in diseases with neurological involvement. Aging & Disease, Vol 5, p. 238.

Edelman S, Craig A, Kidman AD (2000) Can psychotherapy increase the survival time of cancer patients? Journal of Psychosomatic Research, Vol 49, p. 149.

Fontana L, Partridge L (2015) Promoting health and longevity through diet: from model organisms to humans. Cell, Vol 161, p. 106.

Furst AJ, Rabinovici GD, Rostomian AH, Steed T, Alkalay A, Racine C, Miller BL, Jagust WJ (2012) Cognition, glucose metabolism and amyloid burden in Alzheimer's disease. Neurobiology of Aging, Vol 33, p. 215.

Gold PE, Cahill L, Wenk GL (2003) The lowdown on Ginkgo biloba. Scientific American, April, p. 86.

Hall KT, Loscalzo J, Kaptchuk TJ (2015) Genetics and the placebo effect: the placebome. Trends in Molecular Medicine, Vol 21, p. 285.

Kaeberlein M, Rabinovitch PS, Martin GM (2015) Healthy aging: the ultimate preventative medicine. Science, Vol 350, p. 1191.

Lee MS, Pittler MH, Ernst E (2008) Effects of reiki in clinical practice: a systematic review of randomised clinical trials. International Journal of Clinical Practice, Vol 62, p. 947.

Marchalant Y (2009) Cannabinoids attenuate the effects of aging upon neuroinflammation and neurogenesis. Neurobiology of Disease, Vol 34, p. 300.

Markowska AL, Stone WS, Ingram DK, Gold PE, Pontecorvo MJ, Wenk GL, Olton DS (1989) Individual differences in aging: behavioral and neurobiological correlates. Neurobiology of Aging, Vol 10, p. 31.

Montagne A, Barnes SR, Sweeney MD, Halliday MR, Sagare AP, Zhao Z, Toga AW, et al (2015) Blood–brain barrier breakdown in the aging human hippocampus. Neuron, Vol 85, p. 296.

Mosconi L, Mistur R, Switalski R, Brys M, Glodzik L, Rich K, Pirraglia MA et al (2009) Declining brain glucose metabolism in normal individuals with a maternal history of Alzheimer disease. Neurology, Vol 72, p. 513.

Perls T, Levenson R, Regan M, Puca A (2002) What does it take to live to 100? Mechanisms of Ageing and Development, Vol 123, p. 231.

Rammsayer T (1989) Is there a common dopaminergic basis of time perception and reaction time? Neuropsychobiology, Vol 21, p. 37.

Riera CE, Dillin A (2015) Tipping the metabolic scales towards increased longevity in mammals. Nature Cell Biology, Vol 17, p. 196.

Roodenrys S (2002) Chronic effects of Brahmi (Bacopa monnieri) on human memory. Neuropsychopharmacology, Vol 27, p. 279.

Sala SD (Ed.) (1999) Mind Myths: Exploring Popular Assumptions About the Mind and Brain. Malden, MA: John Wiley & Sons.

Science and Technology: Public Attitudes and Public Understanding. Science Fiction and Pseudoscience. Vol 7, p. 21, Washington, DC: National Science Foundation.

Sørensen HJ, Foldager L, Roge R, Pristed SG, Andreasen JT, Nielsen J (2014) An association between autumn birth and clozapine treatment in patients with schizophrenia: a population-based analysis. Nordic Journal of Psychiatry, Vol 68, p. 428.

Stein TD, Alvarez VE, McKee AC (2014) Chronic traumatic encephalopathy: a spectrum of neuropathological changes following repetitive brain trauma in athletes and military personnel. Alzheimers Research & Therapy, Vol 6, p. 4.

Wang Y, Hekimi S (2015) Mitochondrial dysfunction and longevity in animals: untangling the knot. Science, Vol 350, p. 1204.

Wenk GL (1989) An hypothesis on the role of glucose in the mechanism of action of cognitive enhancers. Psychopharmacology, Vol 99, p. 431.

Wenk GL, Olton DS, Hughey D, Engisch KL (1987) Animal models of cholinergic dysfunction related to aging and senile dementia. Gerontology, Vol 33, p. 277.

Wenk GL, Hauss-Wegrzyniak, Willard L (2000) Pathological and biochemical studies of chronic neuroinflammation may lead to therapies for Alzheimer's Disease. In: P Patterson, C Kordon, & Y Christen (Eds.), Research and Perspectives in Neurosciences: Neuro-Immune Neurodegenerative and Psychiatric Disorders and Neural Injury, p. 73. Heidelberg: Springer-Verlag.

Wenk GL (1989) Nutrition—cognition and memory. In: RB Weg (Ed.), Topics in Geriatric Rehabilitation, Vol 6: Nutrition and Rehabilitation, p. 79. Rockville, MD: Aspen Publishers.

Wenk GL, Olton DS (1989) Cognitive enhancers: potential strategies and experimental results. Progress in Neuro-Psychopharmacology & Biological Psychiatry, Vol 13, p. S117.

Wenk GL (1988) Amnesia and Alzheimer's disease: Which neurotransmitter system is responsible? Neurobiology of Aging, Vol 9, p. 640.

Chapter 6

Bayatti N, Moss JA, Sun L, Ambrose P, Ward JF, Lindsay S, Clowry GJ (2008) A molecular neuroanatomical study of the developing human neocortex from 8 to 17 postconceptional weeks revealing the early differentiation of the subplate and subventricular zone. Cerebral Cortex, Vol 18, p. 1536.

Bondy ML, Scheurer ME, Malmer B, Barnholtz-Sloan JS, Davis FG, Il'yasova D, Kruchko C, et al (2008) Brain tumor epidemiology: Consensus from the brain tumor epidemiology consortium (BTEC). Cancer, Vol 113, p. 1953.

Borrell V, Reillo I. 2012. Emerging roles of neural stem cells in cerebral cortex development and evolution. Developmental Neurobiology, Vol 72, p. 955.

Carlson NR. 2012. Physiology of Behavior, 11th ed. Essex, UK: Pearson.

Cheung AF, Kondo S, Abdel-Mannan O, Chodroff RA, Sirey TM, Bluy LE, . . . Molnár Z (2010) The subventricular zone is the developmental milestone of a 6-layered neocortex: Comparisons in metatherian and eutherian mammals. Cerebral Cortex, Vol 20, p. 1071.

Cheung AF, Pollen AA, Tavare A, DeProto J, Molnar Z (2007) Comparative aspects of cortical neurogenesis in vertebrates. Journal of Anatomy, Vol 211, p. 164.

Defelipe J (2011) The evolution of the brain, the human nature of cortical circuits, and intellectual creativity. Frontiers in Neuroanatomy, Vol 5, p. 29.

Dehay C, Kennedy H (2007) Cell-cycle control and cortical development. Nature Review Neuroscience, Vol 8, p. 438.

Dorus S, Vallender EJ, Evans PD, Anderson JR, Gilbert SL, Mahowald M, . . . Lahn BT (2004) Accelerated evolution of nervous system genes in the origin of Homo sapiens. Cell, Vol 119, p. 1027.

Fish JL, Dehay C, Kennedy H, Huttner WB (2008) Making bigger brains—the evolution of neural-progenitor-cell division. Journal of Cell Science, Vol 121, p. 2783.

Fujiwara H, Hirao K, Namiki C, Yamada M, Shimizu M, Fukuyama H, Hayashi T et al (2007) Anterior cingulate pathology and social cognition in schizophrenia: A study of gray matter, white matter and sulcal morphometry. NeuroImage, Vol 36, p. 1236.

Hecht D (2010) Depression and the hyperactive right-hemisphere. Neuroscience Research, Vol 68, p. 77.

Ingalhalikar M, Smith A, Parker D, Satterthwaite TD, Elliott MA, Ruparel K, Hakonarson H, et al (2014) Sex differences in the structural connectome of the human brain. Proceedings of the National Academy of Sciences, USA, Vol 111, p. 823.

Kandel ER, Schwartz JH, Jessell TM, Siegelbaum SA, Hudspeth AJ (2012) Principles of Neural Science, 5th ed. New York: McGraw-Hill.

Kriegstein A, Alvarez-Buylla A (2009) The glial nature of embryonic and adult neural stem cells. Annual Review of Neuroscience, Vol 32, p. 149.

Långsjö JW, Alkire MT, Kaskinoro K, Hayama H, Maksimow A, Kaisti KK, Aalto S, et al (2012) Returning from oblivion: imaging the neural core of consciousness. Journal of Neuroscience, Vol 32, p. 4935.

Lepousez G, Nissant A, Lledo P-M (2015) Adult neurogenesis and the future of the rejuvenating brain circuits. Neuron, Vol 86, p. 387.

Marchalant Y, Brothers HM, Norman GJ, Karelina K, DeVries AC, Wenk GL (2009) Cannabinoids attenuate the effects of aging upon neuroinflammation and neurogenesis. Neurobiology of Disease, Vol 34, p. 300.

Marchalant Y, Brothers HM, Wenk GL (2009) Cannabinoid agonist WIN-55,212-2 partially restores neurogenesis in the aged rat brain. Molecular Psychiatry, Vol 14, p. 1068.

Molnar Z. 2011. Evolution of cerebral cortical development. Brain Behavior Evolution Vol 78, p. 94.

Mota B, Herculano-Houzel S (2015) Cortical folding scales universally with surface area and thickness, not number of neurons. Science, Vol 349, p. 74.

Pearson J, Westbrook F (2015) Phantom perception: voluntary and involuntary nonretinal vision. Trends in Cognitive Sciences, Vol 19, p. 278.

Rakic P (2009) Evolution of the neocortex: A perspective from developmental biology. Nature Reviews Neuroscience, Vol 10, p. 724.

Saladin K (2007) Anatomy and Physiology: The Unity of Form and Function. New York: McGraw Hill.

Striedter GF, Srinivasan S, Monuki ES (2015) Cortical Folding: When, Where, How, and Why? Annual Review Neuroscience, Vol 38, p. 291.

Sun T, Hevner RF (2014) Growth and folding of the mammalian cerebral cortex: from molecules to malformations. Nature Reviews Neuroscience, Vol 15, p. 217.

Tremblay M, Stevens B, Sierra A, Wake H, Bessis A, Nimmerjahn A (2011) The role of microglia in the healthy brain. Journal of Neuroscience, Vol 31, p. 16064.

Wenk GL (2015) Your Brain on Food: How Chemicals Control Your Thoughts and Feelings, 2nd ed. Oxford: Oxford University Press.

Wijdenes LO, Marshall L, Bays PM (2015) Evidence for optimal integration of visual feature representations across saccades, Journal of Neuroscience, Vol 35, p. 10146.

INDEX